SPACE SCIENCE, EXPLORATION AND POLICIES

RECENT ADVANCES IN SATELLITE RESEARCH AND DEVELOPMENT

SPACE SCIENCE, EXPLORATION AND POLICIES

Additional books in this series can be found on Nova's website under the Series tab.

Additional e-books in this series can be found on Nova's website under the e-book tab.

COMPUTER SCIENCE, TECHNOLOGY AND APPLICATIONS

Additional books in this series can be found on Nova's website under the Series tab.

Additional e-books in this series can be found on Nova's website under the e-book tab.

SPACE SCIENCE, EXPLORATION AND POLICIES

RECENT ADVANCES IN SATELLITE RESEARCH AND DEVELOPMENT

SAMUEL GARDINER
AND
KENNETH P. OLSEN
EDITORS

nova publishers
New York

For permission to use material from this book please contact us:
Telephone 631-231-7269; Fax 631-231-8175
Web Site: http://www.novapublishers.com

Library of Congress Cataloging-in-Publication Data

ISBN: 978-1-62417-443-8

Published by Nova Science Publishers, Inc. † New York

CONTENTS

PREFACE

In this book, the authors present new satellite development research. Topics discussed include satellite-linked GPS collars deployed on elephants in Tsavo ecosystems in Kenya; enhancing resiliency of broadband satellite communication emergency networks through propagation impairment mitigation techniques; satellite applications for very short range weather forecasting systems in Southern African developing countries; eliminating ionospheric effects in multifrequency global navigation satellite systems; attitude control systems for arc-second stabilization of 30-kg micro astronomy satellites; and third generation GPS systems.

Chapter 1 – The authors aimed to deploy satellite-linked GPS collars to ten elephants at different locations at Tsavo East and West National Parks. However, nine elephants (3 males and 1 female in Tsavo East and 3 males and 2 females in Tsavo West) were deployed with the collars as one collar malfunctioned. The movement patterns of the collared elephants will be monitored for the next 20 months. This operation was carried out by International Fund for Animal Welfare staff and Kenya Wildlife Service scientists, wardens', veterinary doctor, and capture rangers. The preliminary preparations involved having planning meetings to identify the personnel, collaring operation sites, equipments, and materials to be used during the operation. A final planning meeting was held at Voi Wildlife Lodge on 28 February 2012. The meeting was to ensure that all personnel, materials, and equipment were available. During this meeting, teams to be involved in the operation were constituted, a brief on the collaring operation process was provided, safety/security issues were discussed, and operations starting time was agreed upon. During the operation, daily debriefing meetings were held to review the day's activities and address issues that might have arisen. Each morning, the operation started at around 5.30a.m with the ground team departing to designated site. The light aircraft and helicopter followed at 6:00a.m and 6:15a.m respectively. The operation involved three teams (i.e., the search, darting, and ground teams). The ground team was the first to depart to the potential areas. They stayed within that general potential area awaiting directions from the search team. The search team was then next to fly early in the morning with a light aircraft to the potential site. The team searched for potential groups group at agreed localities (Sangayaya, Ithumba, Ndia Ndaasa, and Emusaya in TENP; and Maktau, Kasigau, Jipe Njukini and Kamboyo in TWNP). Upon spotting a potential group of elephants at agreed site, the search team called in the darting team. The search team then led the darting team to the group. The darting team aged and sexed an identified elephant for darting from the group. Then upon agreeing on the individual, the veterinary doctor darted the animal. Darting was done from a helicopter using a Dan inject gun. The immobilizer and revival drugs used were etorphine hydrochloride and diprenorphine at the dosage of 16-18mg and 60mg respectively. The ground team was called in and directed to the immobilized elephant by the search team. By the end of the operation, satellite-linked GPS collars were deployed to 6 male elephants (3 each in both Tsavos) and three female elephant (one at TENP and two at TWNP). The age of the collared elephants ranged from 17-40 years. The collared elephants will be monitoring for the next 20 months. The information generated will be used to understand the movement ecology and space use of elephants in Tsavo East and West National Parks as well as map their migratory corridors and dispersal area.

Chapter 2 – In this chapter the main issues raised from the employment of telecommunications satellites at emergency communications operating at high frequencies are described. First, the need of telecommunication systems in emergency situations is analyzed showing that these are mandatory for the disaster relief. More specifically, the chapter is focused on the presentation of the various telecommunication services which must be supported, as well as for the backhauling of these networks, by satellite systems. Since, broadband satellite communication systems operate mainly above 10GHz, the propagation impairments at these frequencies are described. They depend mainly on precipitation, atmospheric gases, clouds and tropospheric turbulence. Apart from models for the prediction of the first-order statistical characteristics, models for the generation of time series are also presented. Great attention is paid on rain attenuation, since it is the dominant fading mechanism for the satellite communication systems. Moreover, the most important propagation impairment mitigation techniques that have been proposed in the literature are presented in detail. Finally, general concepts for future research are given and some conclusions are derived.

Chapter 3 – The strength of nowcasting and very short-range forecasting (up to 12 hours lead time) lies in the fact that it provides location-specific forecasts regarding the initiation, growth, movement and dissipation of weather phenomena. In an ideal world, weather radar systems are the most important part of nowcasting. The harsh reality, however, is that many developing countries (DC), and even more so Least Developing Countries (LDC), do not have operational radar systems at all. Furthermore, the countries which are fortunate enough to have radar systems, are struggling to maintain and sustain these powerful data sources. Nowcasting and very short-range forecasting systems for developing countries and/or least developing countries should thus be focusing on using tools such as satellite data and other observation data to address these needs.

The usefulness of Meteosat Second Generation (MSG) geostationary satellite products for nowcasting purposes goes far beyond the availability of 12 channels which can be visualized every 15 minutes. It also includes satellite-based instability indices which aide in the very short-range forecasting of convection as well as other products supplied by the Nowcasting Satellite Application Facility (SAFNWC). Satellite-based precipitation estimations find very valuable applications in the World Meteorology Organization's (WMO) Flash Flood Guidance System (FFGS) projects in various developing regions of the world. The FFGS is a hydrometeorological modeling system that predicts the potential for flash floods based on hydrological conditions that may lead to flooding in a specific basin. The Southern African Regional Flash Flood Guidance (SARFFG) project is one of the sub-regional projects of the WMO's global FFGS program. SARFFG is highly dependent on satellite-based precipitation estimations to determine whether small catchments have the potential to experience flash floods.

The WMO Severe Weather Forecasting Demonstration Projects (SWFDP) aim to improve forecasting and severe weather forecasting in DC and LDC by providing access to forecasting information through a website. Feedback from the participating National Meteorological and Hydrological Services in the Southern African SWFDP indicates that the SWFDP has assisted to improve warning services in many countries. At the fourth meeting of the WMO Commission for Basic Systems SWFDP Steering Group in Geneva, February 2012, it was recognized that one of the main challenges for the SWFDP was "the need for very short-range forecasting tools, to address especially the rapid onset of localized severe thunderstorms which can produce heavy precipitation and strong wind, given the absence of adequate real-time observational networks, especially weather radar coverage." It was also agreed that real-time satellite rainfall estimates have proven particularly useful in regions where radar coverage and rain gauges are sparse.

In this chapter an overview and examples will be provided of recent applications and products which were developed using geostationary satellite data to improve nowcasting and very short-range forecasting of convection, precipitation and flash floods in areas where other remote sensing tools are not readily and/or operationally available.

Chapter 4 – Methods for environmental diagnostics with the use of global navigation satellite systems (GNSS) have been progressing intensively. Failures of these methods are largely caused by the anisotropic, space-time-variable ionospheric plasma that is in the path of a satellite signal. The ionospheric effects are usually removed using dispersive features of the ionospheric plasma. To do it, phase or/and time delays of a satellite signal is measured at two frequencies. The dual-frequency measurements neutralize the major part of the ionospheric error (the first-order error). But this accuracy is already insufficient for high-precision measurements needed for monitoring the lower atmosphere to forecast the weather and climatic changes and for examining the deformation of the Earth's surface to predict earthquakes and volcanic activity. This chapter summarizes findings of the investigations into the residual ionospheric error. Here the authors mainly rely on authors' publication supplemented by literature data and some unpublished findings. In the geometrical optics approximation, the errors of phase and group paths of an ionospheric radio wave at GNSS frequencies may be represented as series in inverse powers of frequency. It is possible to avoid the commonly encountered inaccuracies here, if wave propagation of certain (right) polarization is considered instead of ordinary or extraordinary waves as in traditional investigations. The authors show how the dual-frequency measurements can be exploited to remove not only the first-order error, but also the second-order error associated with the geomagnetic effect on the refractive index in the ionospheric plasma. After eliminating the second-order error in dual-frequency measurements, the authors can remove third-order effects associated with ray bending in the inhomogeneous ionosphere, using triple-frequency measurements. It is shown that in addition to the background ionosphere a considerable (often dominating) contribution to this error is made by random ionospheric inhomogeneities. Effects of ionospheric inhomogeneities in triple-frequency GNSS can be eliminated only for inhomogeneities with transversal scales exceeding the Fresnel radius. Diffraction makes it impossible to remove the effects of inhomogeneities with smaller scales. This leads to the residual error in triple-frequency measurements. The authors discuss possibilities of existing and currently developed spatial signal processing methods being able to eliminate diffraction effects and improve accuracy of multifrequency GNSS measurements.

Chapter 5 – This study investigated the spatial-temporal patterns of elephant poaching in Tsavo East National Park to establish and map the density of poached elephants as well as model elephant poaching hotspots. Data on poached elephants for the period between 10 October 1989 and 2 July 2005 was acquired from the Kenya Wildlife Service elephant mortality database at the headquarters. The results indicate a higher concentration of overall poaching mainly in the central and northern parts of the park. Both dry and wet season elephant poaching ellipses indicate that seasonal elephant poaching exhibited similar spatial orientation and their centers of distribution lay approximately in the central part of the park. Kernel density analysis depicted dry season poaching hotspots on the Tiva and Galana rivers in the northwest and central-western parts of the park. Sites with moderate dry season elephant poaching included the southeastern tip of the park on Voi River and the central-eastern part. There were two main wet season poaching hotspots occurring along the lower part of Galana River across the park. Moderate wet season poaching sites were spread along Tiva River to the north of the park through the central parts of the park, to the southeastern tip on Voi River. The risk to elephant poaching map for wet season indicated areas at highest risk as occurring mainly in the central-east, south of Sala gate toward the southeastern edge of park, and in the northeastern parts of the park. The area of highest risk to elephant poaching during the dry season occurred along the Tiva River, central parts of the Galana and Voi rivers. Elephant risk to poaching during the dry season indicated areas of highest risk as concentrated along the main rivers, where water is available throughout the year. An overall risk to poaching map indicated areas with the highest risk to elephant poaching occurred in the central part of the Tiva River and the central-eastern parts of the Galana River. In order to provide adequate security of elephants throughout the year, there is need to re-deploy wildlife policing resources (equipment and personnel) to high risk areas as well as have temporally security mobile units in the open plains during the wet season. There is also need to open more security patrol roads on the eastern side of Sangayaya and Ithumba areas.

Chapter 6 – This chapter focuses on the development of a precise attitude control system for a 35 kg small astrometry satellite. In general, conventional methods used for an attitude control in standard-sized satellites are not applicable to small satellites because of constraints of power consumption, space requirements, and mass. In addition, small satellites are more greatly affected than standard-sized satellites by attitude disturbances because of their relatively small moment of inertia. Although a magnetic disturbance torque is considered negligible in standard-sized satellites, it is dominant in small satellites. To realize precise attitude stabilization in small satellites, the satellites should use an attitude control method that satisfies the abovementioned constraints and reduces the effect of magnetic disturbance. This study proposes a new method to solve these problems. Through an example of an astronomical observation mission using small satellites, it is demonstrated that precise attitude control is feasible in small satellites.

Chapter 7 – The third-generation Global Positioning Satellite system (GPS system) is now in the early development phase by Celestia Corporation. The new system, major components of which were designed by the author, has substantial advantages by comparison with the current GPS system. Doppler correction and signal tracking circuitry in the receivers have been eliminated, which resulted in a substantial reduction in the size of the receiver (in the near future, a GPS receiver will fit inside an ordinary wrist watch). Maintenance signals, which are frequently transmitted from the earth stations to the satellites, have also been substantially reduced. Simulations that characterize the User Equivalent Range Errors show the advantages of the new system. Overall, the new system is substantially more reliable, offers better performance, and is substantially cheaper than the current GPS system. The appendix to this chapter gives a closed-form solution for the GPS geolocation equations.

In: Recent Advances in Satellite Research and Development ISBN: 978-1-62417-443-8
Editors: Samuel Gardiner and Kenneth P. Olsen

Chapter 1

SATELLITE-LINKED GPS COLLARS DEPLOYED ON ELEPHANTS IN TSAVO ECOSYSTEM, KENYA

Shadrack Ngene[1], Steve Njumbi[2], Bernard Ngoru[3], Martha Nzisa[3], Elephus Bitok[4], Jeremiah Poghon[3], Shadrack Muya[5] and Alex Gombe[2]

[1]Kenya Wildlife Service, Biodiversity Research and Monitoring, Elephant Programme, Nairobi, Kenya
[2]International Fund for Animal Welfare, Nairobi, Kenya
[3]Kenya Wildlife Service, Tsavo Conservation Area, Kenya
[4]Kenya Wildlife Service, Biodiversity Research and Monitoring, Southern Conservation Area, Nairobi, Kenya
[5]Jomo Kenyatta University of Agriculture and Technology, Nairobi, Kenya

ABSTRACT

We aimed to deploy satellite-linked GPS collars to ten elephants at different locations at Tsavo East and West National Parks. However, nine elephants (3 males and 1 female in Tsavo East and 3 males and 2 females in Tsavo West) were deployed with the collars as one collar malfunctioned. The movement patterns of the collared elephants will be monitored for the next 20 months. This operation was carried out by International Fund for Animal Welfare staff and Kenya Wildlife Service scientists, wardens', veterinary doctor, and capture rangers. The preliminary preparations involved having planning meetings to identify the personnel, collaring operation sites, equipments, and materials to be used during the operation. A final planning meeting was held at Voi Wildlife Lodge on 28 February 2012. The meeting was to ensure that all personnel, materials, and equipment were available. During this meeting, teams to be involved in the operation were constituted, a brief on the collaring operation process was provided, safety/security issues were discussed, and operations starting time was agreed upon. During the operation, daily debriefing meetings were held to review the day's activities and address issues that might have arisen. Each morning, the operation started at around 5.30a.m with the ground team departing to designated site. The light aircraft and helicopter followed at 6:00a.m and 6:15a.m respectively. The operation involved three teams (i.e., the search, darting, and ground teams). The ground team was the first to depart to the potential areas. They stayed within that general potential area awaiting directions from the search team. The search team was then next to fly early in the morning with a light aircraft to the potential site. The team searched for potential groups group at agreed localities (Sangayaya, Ithumba, Ndia Ndaasa, and Emusaya in TENP; and Maktau, Kasigau, Jipe Njukini and Kamboyo in TWNP). Upon spotting a potential group of elephants at agreed site, the search team called in the darting team. The search team then led the darting team to the group. The darting team aged and sexed an identified elephant for darting from the group. Then upon agreeing on the individual, the veterinary doctor darted the animal. Darting was done from a helicopter using a Dan inject gun. The immobilizer and revival drugs used were etorphine hydrochloride and diprenorphine at the dosage of 16-18mg and 60mg respectively. The ground team was called in and directed to the immobilized elephant by the search team. By the end of the operation, satellite-linked GPS collars were deployed to 6 male elephants (3 each in both Tsavos) and three female elephant (one at TENP and two at TWNP). The age of the collared elephants ranged from 17-40 years. The collared elephants will be monitoring for the next 20 months. The information generated will be used to understand the movement ecology and space use of elephants in Tsavo East and West National Parks as well as map their migratory corridors and dispersal area.

Keywords: Elephant, Satellite-linked, GPS collars, Tsavo, Ecosystem

INTRODUCTION

Nine elephants were successfully fitted with satellite-linked GPS tracking collars in Kenya's vast Tsavo East and West National Parks by a combined team of Kenya wildlife Service (KWS) and International Fund for Animal Welfare (IFAW) staff. This brings to twelve the total number of elephants fitted with satellite-linked GPS collars, after the first phase of a similar operation a year ago. This is the first large scale monitoring of elephant movements using satellite technology in the Tsavo East and West National Parks. The last tracking initiative was done with conventional radio tracking collars deployed in 1972, forty years ago. IFAW and KWS fitted these elephants with the collars to give current information on elephant movements and their usage of the Tsavo ecosystem. The aim is to mitigate threats of poaching and human-elephant conflicts witnessed in recent times. The threats are due to increase of human population around the two parks.

Human population has significantly increased in the area since the 1972 tracking project (GoK, 2010). The increase has resulted to habitat fragmentation and encroachment into former elephant rangelands through crop farming and other human infrastructure changes. These changes are a threat to Tsavo's ecological integrity whose habitat has previously, due to its large size, being considered Kenya's most viable natural habitat for long-term survival of its long ranging elephant population. In addition, the fragmentation has led to more cases of human-elephant conflict (HEC) in the ecosystem.

KWS reports at least 278 elephants killed for their ivory in 2011 (KWS, 2012). The purpose of collaring the elephants is to gather information to guide management decisions regarding human-elephant conflict resolution, anti-poaching efforts, and identification of critical corridors and dispersal areas outside the protected areas. Information collected from the five elephants collared last year, two of which have died, already point to two interesting observations. First, that elephants are "area specific,"meaning they have preferred areas in the ecosystem that they stick to a majority of the time. We can infer this behavior as resource partitioning of the ecosystem by the elephants. For management this is good and bad news. Good because managers can better predict where the elephants are and concentrate security and HEC effort in those areas. It is bad because poachers can also predict where to find elephants. Second, one female elephant has spent the majority of her time outside the protected area in a community ranch south of Tsavo East. In terms of ecology and management, this "proves" there is a corridor and dispersal area important to elephants outside the park boundaries that needs to be secured. But it is bad news because security is not as tight outside the park.

KWS and IFAW will continue monitoring the collared elephants for the next 20 months. We report on the March 2012 collaring operation and propose recommendations to guide future collaring operations.

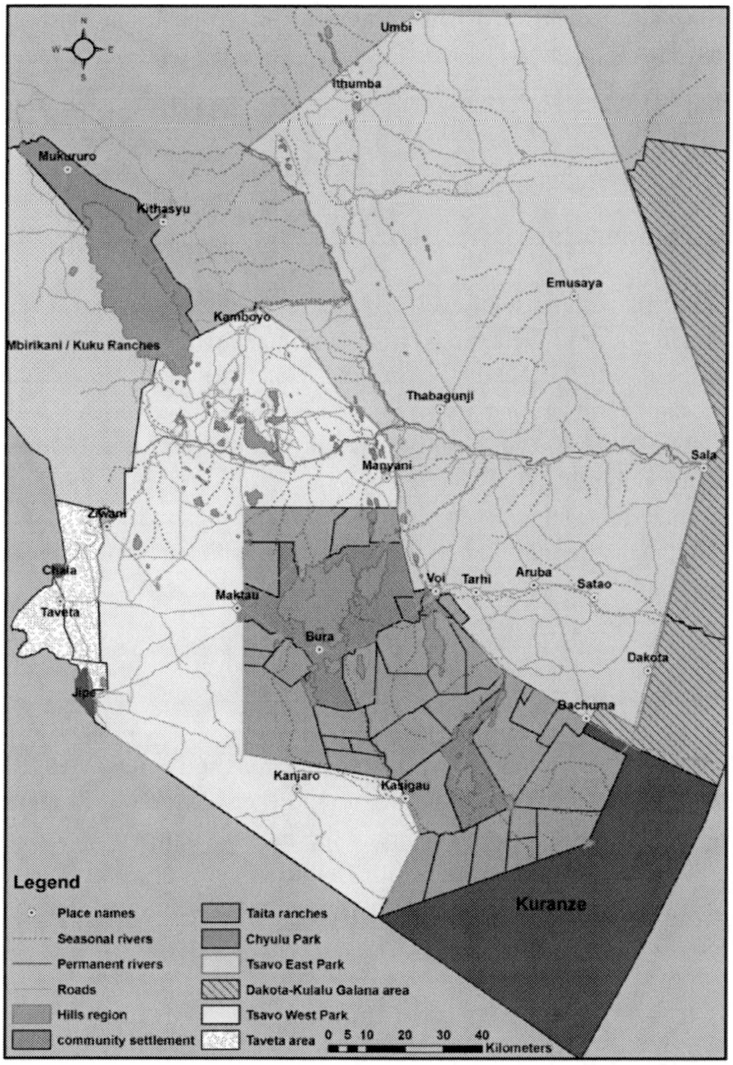

Figure 1. A Map showing the location of the study area (The Tsavo ecosystem).

STUDY AREA

Location

The study area is located in southern Kenya between 2° and 4° S and 37°30′ and 39°30′ E (Figure 1).

The ecosystem covers an area of over 40,000 km², and is subdivided into the following: Tsavo West National Park, measuring about 9,000 km², Tsavo East National Park measuring approximately 13,000 km², Chyullu National Park approximately 780 km^2 and the rest are the surrounding ranches and community land. It borders Mkomazi National park in Tanzania. The three National parks cover 22,780km^2.

Tsavo Conservation Area (TCA) is one of the eight (8) KWS conservation areas and comprises of protected areas, Ranches, Private and community land. It is second largest contiguous protected area/park in Africa, after Kruger NP in South Africa. It comprises the largest protected area in Kenya - Tsavo East, West and Chyulu occupy about 52% of total protected areas in the country and about 3.9% of Kenya's surface area. TCA is also home to largest single elephant population in the country.

Rainfall and Temperature

The climate of the area is semi-arid, with an unpredictable, bimodal rainfall distribution of between 200 and 700 millimeters per annum (Wijngaarden 1985). The long rainy season is experienced in the months of March - April/ May and the short rain season in November - December. The short rains are generally considered to be more reliable and less erratic than the long rains. Rainfall in Tsavo West is generally higher and usually less erratic in spatial and temporal distribution than in Tsavo East (Wijngaarden 1985; Leuthold and Sale, 1973).

Average normal daily temperatures range between 20°C and 30°C. The temperatures are slightly higher in the dry season than in the wet season with a general decrease in temperature with increasing altitude.

Terrain, Geology, Soils

Tsavo Conservation Area has a gentle northwest slope in most areas while a few areas slope to the west and north. The study area ranges between 100-900 meters above the sea level. Most soils have a sandy- loam texture in the sub-soil however in some places they have a sandy texture. Soils of loamy texture are found as deposits in valleys a few places are covered by rocks. The soils exhibit little variations in colour, drainage, porosity and temperatures.

Flora and Fauna

TCA falls along a lowland savanna with vegetation dominated by *Acacia–Commiphora*, in which the density of trees and shrubs varies significantly over time and space (EIA report for Jipe-Rombo fence). Other plant species include *Dobera glabra, Newtonia hildebrandtii, Acacia elatior,* and *Kigelia africana*. Common shrubs include *Azima tetracantha, Capparis sepiaria, Pluchea dioscordis, Salvadora persica, Combretum ukambensis, Cordia goetzei, Gardenia jovis-tonantis, Lawsonia inermis, Vernonia hildebrandtii* and *Meyna tetraphylla*. The main grass species include *Brachiaria deflexa, B. leersoides, Cenchrus ciliaris, Digitaria macroblephara, D. rivae, Latipes senegalensis, Panicum maximum, Aristida adscensionis, Chloris roxburghiana, Tetrapogon tenellus* and *Sporobolus helvolus*.

The common wildlife species in the area are elephants (*Loxodonta africana*), African buffalo (*Syncerus caffer*), Burchell's zebra (*Equus burchellii*), Coke's hartebeest (*Alcelaphus buselaphus*), waterbuck (*Kobus ellipsiprimnus*), Grant's gazelle (*Gazella grantii*), Impala (*Aepyceros melampus*), hippopotamus (*Hippopotamus amphibius*), Giraffe (*Giraffa camelopardalis*), lesser kudu (*Tragelaphus imberbis*), gerenuk (*Litocranius walleri*), warthog (*Phacochoerus aethiopicus*), fringe-eared oryx (*Oryx beisa callotis*)and eland (*Taurotragus oryx*). Common domestic animals kept on the surrounding areas include, indiginous cattle (*Bos indicus*), sheep (*Ovis aries*), goats (*Capra hirtus*), and donkeys (*Equus asinus*). Large carnivore species found in the area include lions (*Panthera leo*), spotted hyena (*Crocuta crocuta*), cheetah (*Acinonyx jubatus*), leopard (*Panthera pardus*), and black-backed jackal (*Canis mesomelas*).

Land Use

Within the protected areas, wildlife viewing based tourism tourism is the main type of land use. Outside the protected areas, the communities practice nomadic pastoralism and sendentary livestock keeping. Commercial livestock keeping occurs especially in the ranches (Kulalu and Galana among others). Small scale cropping mixed with livestock is dominant outside the protected area networks. Large scale crop farming is limited to the west and central parts of the ecosytem. There is also intensive horticultural crop farming practiced in the areas around Ziwani and Challa. Overstocking of livestock has led to frequent incursions into the protected areas. Most of the livestock are found in private ranches adjacent to the parks as well as in individual farms. The ranches also act as wildlife dispersal areas.

Elephant Population Status and Trend in Tsavo Ecosystem

The Tsavo ecosystem hosts about 12,573 elephants (Ngene et al., 2011). Based on detailed sample and total aerial counts over the years, the elephants population has been fluctuating (Blanc et al., 2007; Ngene et al., 2011). Recent total and sample aerial counts are those by Olindo et al. (1988), Douglas-Hamilton et al. (1994), Kahumbu et al. (1999), Omondi et al. (2002), Omondi and Bitok (2005), Omondi et al. (2008), Ngene et al. (2011) and by Cobb (1976); Leuthold (1976); WCMD (1976); IUCN (1978), and Inamdar (1996). The 1988 counts showed a 75% and 87% decline in elephant numbers within the protected areas and adjacent non-protected areas respectively since the 1972 total counts (Olindo et al., 1988). Since 1988, there has been a steady increase in elephant numbers, up from the steep decline in the late 1970s and early 1980s. The distribution of elephants in 1989 confirmed that elephants previously counted along the periphery had moved further inside the parks. Despite this, the 1989 count was the first authoritative confirmation that the elephant population was on a recovery course, a trend observed till 2011. The count revealed that the Tsavo population had increased by 5% since 1999 (1999 = 8,068 elephants; 2002 = 9,284 elephants; Kahumbu et al., 1999). This population increased to about 10,397 in 2005; 11,733 in 2008 (Omondi et al., 2008); and 12,573 in 2011 (Ngene et al., 2011). There population increased by only 25 between 2008 and 2011 (Ngene et al., 2011).

Materials And Methods

The collaring operation was divided into three phases including the pre-collaring operation planning, actual collaring operation and the debriefing meetings.

The Pre-Collaring Operation Planning

The collaring exercise was preceded with extensive planning to ensure that the event was successful and that all possible incidents were considered. Two major planning meetings were organized at Voi Wildlife lodge comprising of all the targeted participants in the exercise. The first meeting took place on 18 January followed by 28 February 2012. The importance of these meetings was to ensure that all the participants were conversant with the process, the operation teams to bond and assign roles and responsibilities to respective members, and ensure that all the equipment and personnel required for the operation were procured before the D-day and made available to the various teams.

During the meetings, several presentations were made, action plans and budgets developed, timelines set and key teams constituted. Emphasis was made on the importance and the goals of the exercise, and the need to go through all the collaring procedures to avoid making mistakes and getting all details correct for the entire team. Presentation on the purpose, objectives and collaring sites was made followed by the collaring operation, techniques and deployment and collaring teams, chain of command, the do's and don'ts during the operation (Annex 1 and daily operational program (Annex 2) was thoroughly discussed in the meetings. The participants were finally divided into three major teams each with a team leader. These were the spotter team, chopper team and the ground team (Annex 7). Effective communication was also emphasized to ensure that the teams operated in a synchronized manner and all vehicles and aircrafts were equipped with air band radios, mobile radios and or hand held radios.

Actual Collaring Operation

Operation Teams

Three teams were involved in the actual collaring operation. These were:

i. *The search team:* A spotter team identified potential candidate herd for collaring from a husky light aircraft and communicated the information to the darting team in the helicopter who then directed the ground teams (Animal welfare, collar deployment, data collection and security team) to move closer to the site. The team also monitored the movement of the darted animal and advised on the security in general in the vicinity of the darted animal.

ii. *The darting team:* This team was composed of a pilot, a veterinary doctor, two wildlife biologists (one each from IFAW and KWS), who identified the target individual for collaring and directed the veterinary doctor to dart the animal, a camera man, and a security person. The chopper pilot then maneuvered to isolate the identified individual from the rest of the group and allowed the vet to immobilize the animal with a dose of etorphine from a gun propelled syringe. The team was also responsible of driving the darted animal to safety and securing the elephant once fallen. The team carried the collar and other accessories required to deploy it onto the immobilized animal.

iii. *The ground team:* The team consisted of various support units whose role was collar deployment, provision of security, data and sample collection, media, first aid, logistics, transport and communication. This team backed up the darting team to ensure that the elephant was laying in a non-lethal position, the elephant body was cooled by pouring water on its body and ear region, the collar was efficiently and effectively deployed, all required data was collected, the area was secure from threats posed by other elephant groups or other wildlife, and all the persons involved were secure once the immobilized elephant was revived.

The Collaring Operation

The actual collaring operation involved the following in order of approach:

- Locating a target group of elephants with a husky aircraft. This was done by taking off from an airstrip and flying to an agreed upon area in the park where an elephant was to be darted. On arriving at the general area, the husky aircraft searched for potential target groups. The pilot and the researcher agreed upon the group from which an individual will be darted. Then, the helicopter, carrying the darting team was called in. The helicopter was directed to the target group by the team in the husky. Once the darting team located the target, the husky pulled off and started directing the ground team towards the darting team.
- Preparing a dart gun (Dan Inject) for the operation. This was done by the veterinary officer and his assistants before boarding the chopper responsible for the darting to immobilize the selected elephant.

- An individual from the group was identified and darted by the darting team using standard procedures as outlined by Kangwana (1996). These include ensuring that the age of females and males is at least 20 and 25 yrs respectively and that the females do not have a baby of less than two years old. The darted individual was then followed until it becomes immobilized. Darting with etorphine hydrochloride drug at a dosage of 16-18mg was done from the helicopter using a Dan inject gun.
- Once the darted individual fell down, the helicopter landed near the target. The ground team comprising of the collar deployment group, security team, data collection team was then led by the team in the husky aircraft to the location of the immobilized elephant. The spatial arrangement of collaring personnel at recumbent site was as illustrated in figure 2 below. The darting and ground teams then deployed the collar around the neck of immobilized animal starting with securing the elephant by ensuring the elephant is lying in the correct position followed by measuring the neck circumference giving an allowance of 6 inches. Collar measurements were then taken and cut as appropriate followed by pulling the collar underneath the neck using a hooked wire, attaching the lead Weight using the plates, nuts and bolts and ensuring the counter weight is well balanced with the upper unit. Other data collected on the immobilized animal included: blood samples, length and width of the fore and hind and front legs; tail length, diameter of the ear, and samples of ticks on the body parts and tail hairs. Data on the time the animal was darted, the time it fell down, the time the reviver was injected, and the time the animal stood up was also be recorded. During this time, the search team scanned the area to monitor other nearby elephant groups and other wildlife to ensure the darting and ground teams were secure. The logistics team made sure that all the teams had adequate drinking water, packed lunch while the transport team and aircraft attendants ensured that all vehicles were and aircrafts serviceable respectively. First aid team was also on standby in case of injuries.
- Once the collar (plate 1) was deployed and all the samples and data were collected, the revival drug diprenorphine at a dosage of 60mg was injected to the immobilized elephant.
- The team then retreated to the helicopter and vehicles to prepare to take off or drive off once the revived elephant stood up and started to move. The search team monitoring the situation on the ground and kept on communicating with the ground teams regarding safety issues. Thereafter, the search team retreated to another area to search for another target group in another locality or within the same area.
- The same process was followed until all the 9 collars were deployed to 9 different elephants in different locations in Tsavo East National Park.

Debriefing Meetings

This involved holding daily meetings from 18:00 hours to 19:00 hours to discuss progress of the exercise. The meetings which were held at the end of the days' work, was aimed at reviewing the day's operations with a view of making adjustments to ensure success during the following day's operation.

The meetings highlighted the expected roles and responsibilities of the team members and leaders for the following day, timelines and meeting points while at the same time amending the weaknesses of the day's operation. Participants who exhibited excellent performance during the day's operation were acknowledged during the meeting. In addition, the team leaders ensured that they had all that was required for the following day and if anything was missing, it was raised during the meeting.

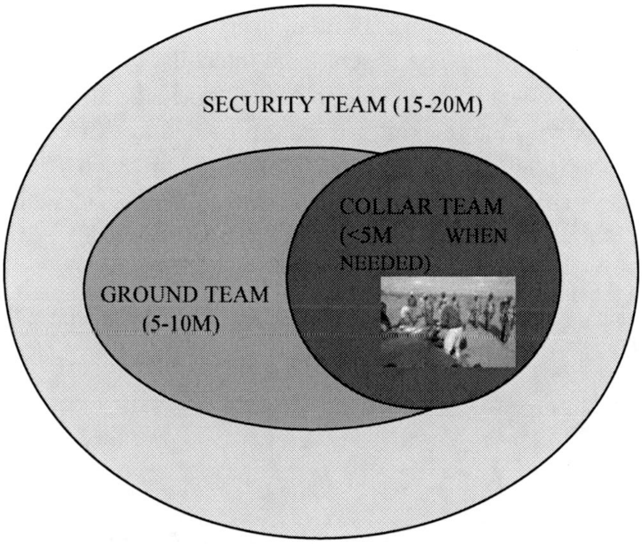

Figure 2. Spatial arrangement of collaring personnel at recumbent site.

All participants were asked to adhere to observe and acquaint themselves to safety rules, to note the chain of command (within/between) and for the pilots, vets, biologists, to adhere to highest standards and techniques.

Chain of Command

The collaring exercise was treated as an operation with a clear chain of command. The Institutional spokespersons and officials in charge of entire operation were the Kenya Wildlife Service (KWS) assistant director in-charge of Tsavo conservation area and the regional director, International fund for animal welfare (IFAW) assisted by the senior Wardens for Tsavo east and west. This was the highest decision making team in the operation. Second in command were the technical team leaders with KWS spotting biologist taking charge of all spotting concerns, the IFAW biologist taking charge of the chopper team, capture warden was the overall team leader in charge of the ground team while the veterinary officer took care of veterinary problems. The ground team had ground Unit Team Leaders comprising of collar deployment, security team leaders consisting of company commanders for Tsavo west and east, data collection, transport management, radio communication, administration and media.

Type of Collars and Data Down Loading

Standard equipments and materials were used in collaring exercise. The collars (plate1) utilize the GPS system for position acquisition and transmit the information via a satellite link. In addition to the GPS positioning the collars emits a VHF beacon for on ground or aerial tracking. The theoretical battery life expectancy is almost 24 months with a one hour data recording schedule. All collars were set to collect positions every hour and to emit a VHF beacon. GPS positions were transmitted every hour while the time label on the positioning data is local time. The data is sent from the collar via a satellite link in batches of one position per hour. This data is accessible from the homepage after logging on as user. The data uploads automatically to the tracking software provided by African Wildlife Tracking of South Africa, when connected to the web and running the program. The downloaded data is exported and saved as an excel spreadsheet. This data can then be added to ARCGIS software for further analysis.

Plate 1. Type of Collar used.

Plate 2. Helicopter ad Husky used.

(3) The Ground Team(s)
Collar deployment: Security: Data Collection: Transport Management: Radio Communication: Media & Communications: Logistics & Administration: First Aid: Aircraft Attendants

Elephant Immobilization and Handling

Elephant immobilization was a critical part of the collaring operation. The need to restrain the animal in the least stress levels and assured security of both personnel and that of the animal was of utmost importance. The drugs used in immobilizing the elephants did not meet all the specifications required and choice of the best alternative was made before embarking on the exercise to avoid adverse effects of immobilization drugs which include respiratory failure, hypothermia/hypothermia and cardiac arrest. Side effects were countered by use of respiratory stimulants and timing of the operation to cool hours of the day. The use of Dan-inject darting system in the safety of a Helicopter was also ideal. Chopper herding of the darted Elephant to safe areas greatly improved the efficiency of the operation. In total nine elephants were safely immobilized and collared in an operation that took three days. An experienced capture team sourced from Nairobi came in handy in handling immobilized Elephant.

The operation was scheduled for cool hours in the morning and late afternoon in order to minimize complications from high temperatures that prevailed in the Tsavo region. 3 cc darts filled with 17 mgs of etorphine and 1000 i.u of hyalase and 18 mgs and 1000 i.u of hyalase were used to immobilize cows and bulls respectively. The darts were propelled by a gas operated dan-inject dart gun from the safety of a Helicopter. The Helicopter was also used to herd the darted elephant close to the roads and away from thick bushes and bad terrain in addition to provision of security to the ground team. After the Elephant fell down it was immediately secured to a good position by the ground team which composed of a vet doctor and capture staff. Various physiological parameters were monitored and appropriate treatment instituted. Collaring was done quickly and took between 9-25 minutes depending on specific challenges. Blood and parasitic samples were taken for lab analysis. Revival was done using an opiod antagonist diprenorphine at three times the etorphine dose. Close monitoring was done at a distance until the Elephant was on its feet and walking before the team left the area.

RESULTS

Proposed and Actual Collaring Sites

Figure 3 below shows the proposed and actual collaring sites in Tsavo East and West National Parks respectively. Nine elephants were collared instead of the planned ten as one of the collar was not functioning. The nine elephants were collared in three days (19-21/03/2012).

The team managed to deploy collars to four elephants each during the first two days whereas on the last day, one collar was deployed to an elephant. On 19 March 2012, the collars were deployed on elephants at Emusaya, Ndia Ndaasa, Ithumba and Sangayaya areas in Tsavo East National Park. On 20 March 2012, collars were deployed to four elephants at Kasigau, Maktau, Jipe, and Njukini areas of Tsavo West National Park. On the last day, a collar was deployed to a male at Kamboyo area of Tsavo West National Park.

Sex, Age, and Drug Dosages of Darted Elephants

Table 1 below presents some of the data collected on the collared elephants, which includes the drugs used during immobilization and revival as well as the specific dosages, the darting range and time effort in collar deployment. The average age of the collared elephants was 22 years for females and 30 years for males. The average darting range was 27 metres from a Chopper aircraft. The average collar deployment time was 13 minutes however, the longest time spent in collar deployment was 25 minutes for the Ndia Ndasa male where extra time was used in treating the severely speared individual.

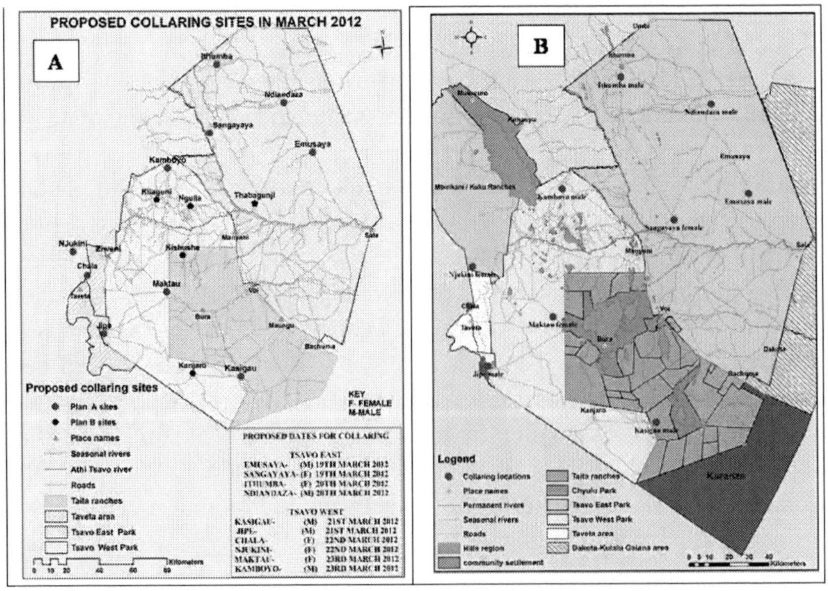

Figure 3. Maps showing collaring operation sites. A: The proposed sites; B: the actual sites where elephants were collared.

Description of Collared Elephants

Description of collared elephants was recorded by both the ground and the darting team. The targeted individuals were darted from the air. Unique identification (ID) reference names were assigned to each darted individual with name IDs having a combination of two letters that refer to the collaring site name, individuals' sex.

The first individual to be collared was the Emusaya Male (EM) which was a male of about 30 years old, with only the left tusk present a feature that would become a surprisingly common occurrence during the exercise.

The second individual to be collared was the Ndia ndasa Male (NM) aged about 35 years and again sporting only a left tusk. The individual was a victim of poaching or possible human wildlife conflict with two arrowheads embedded on the side of his belly. The two veterinary officers in the exercise removed the deeply embedded arrows and treated the elephant before revival. This treatment was necessary as the wounds could become septic since the elephant's thick skin does not allow accumulated pus to seep out which can weaken the animal and eventually cause a slow, painful death.

The third elephant to be collared on the first day was Ithumba Male (IM) estimated to be 23 years old and had only the left tusk. The collaring team had targeted a female individual for this site however safety issues in the closed bush area necessitated quick operation hence it was only realized the collared individual was a young male after the collar was deployed. An extra dose of etorphine was given as the animal showed signs of consciousness within a few minutes into collar deployment.

The last individual to be collared on the first day was the Sangayaya Female (SGF) estimated to be 23 years old and with both tusks intact. The targeted site for collaring was Thabangunji, however, the nearest herd was sighted at Sangayaya. The old collar that had been fitted to the Emusaya male which may have been poached was deployed on this individual and appeared to be tight.

The second day of collaring operation started at Kasigau where a two tusked 40 year old male (KM) was collared at the Kasigau ranch. Of the individuals collared during the operation this was the oldest and largest in body size. Collaring at Kasigau was followed by Maktau site where a one tusked female (MKF) was collared. This individual was in a group of 23 females and had to be driven from the herd to ensure safety for the collar deployment team.

On the second day of collaring the Jipe Male (JM) aged 37 years old with about six inches of its trunk missing probably as a result of being caught in a wire snare. This entire individual's body skin was heavily keratinized.

The last to be collared on 20 march 2012 was the Njukini Female (NJF) estimated to be 23 years old. This individual was found in a group of 80 outside the south-western boundary of Tsavo West National Park and about 4Km to the Kenya-Tanzania border. Important observations made on the herd by the darting team were that the body and tusk sizes of this herd were relatively smaller compared to other Tsavo elephants. The shorter, thinner and straighter tusks and prominently exposed backbones were characteristically distinctive of the population. These observations were confirmed on the ground during collar deployment of NJF which led to the postulation that this could be a remnant of forest elephants that moves between Tsavo West and Kilimanjaro National Park in Tanzania.

Table 1. Sex, age, and drug dosages of darted elephants in TCA March 2012

Elephant Collaring Site	Elephant ID	Collar frequency	Sex	Age (Years)	Etorphine dosage (mg)	Depronophile dosage (mg)	Hylase dosage (mg)	Darting Range (Metres)	Position of dart	Time taken to go down (Minutes)	Revival time (Minutes)	Collar deployment time (Minutes)
Emusaya	EM I 14	150.590	Male	30	18	54	1000	30	Rump	10	2.10	13.00
Ndia Ndasa	NM I 12	150.110	Male	34	17	51	1000	40	Rump	9	1.50	25.00
Ithumba	IM I 13	150.130	Male	17	17	51	1000	30	Rump	2	-	14.00
Sangayaya	SGF I 8	150.650	Female	23	17	51	1000	30	Pelvis	10	1.53	11.00
Kasigau	KM I 17	150.930	Male	40	18	54	1000	20	Rump	10	2.26	9.20
Maktau	MKF I 15	150.890	Female	21	17	51	1000	25	Rump	8	1.50	9.09
Jipe	JM I 9	150.040	Male	37	18	54	1000	20	Rump	10	3.23	12.38
Njukuni	NJF I 16	150.910	Female	23	18	54	1000	30	Rump	4	2.01	10.00
Kamboyo	KMMI 10	150.060	Male	23	18	54	1000	20	Rump	7	2.06	9.29

NB: Elephant revival time refers to the time taken by the elephant to wake up after injection of the revival drug, while collar deployment time is time taken from collar measurements to revival time.

Table 2. Table showing lengths of the legs and those of spoor diameters of different elephants measured during collaring (lengths from the front legs of the elephants). ELE=Elephant; ID=Identification; DIAM=Diameter; LEN=Length; CIR=Circumference; NM=Not Measured; NT=No Tusk; EM=emusaya Male; NM=Ndia Ndasa Male; IM=Ithumba Male; SM=Sangayaya Male; Kasigau Male; MF=Maktau Female; JM=Jipe Male; NJF=Njukini Female; KMM= Kamboyo Male

Date	Ele ID	Front legs information (cm)				Hind legs information (cm)				Other biophysical information (cm)							
		Spoor Diam L	Spoor Diam R	Len L	Len R	Spoor Diam L	Spoor Diam R	Len L	Length R	Tail Len	Ear Diam	Back Len	Shoulder height	Tusk Cir. L	Tusk Cir. R	Tusk Len L	Tusk Len R
19/3	EM	47	50	NM	164	NM	55	NM	156	102	71	350	350	41	NM	101	NM
19 /3	NM	52	52	NM	NM	53	53	133	130	140	161	333	335	NM	42	NM	148
19/3	IM	38	37	NM	128	38	37	NM	121	108	111	265	258	23	NT	65	NT
19 /3	SF	38	38	NM	110	43	43	NM	NM	121	137	285	270	NM	22	80	80
20/3	KM	52	50	NM	NM	58	58	NM	165	142	154	330	420	NM	44	131	105
20/3	MF	38	37	NM	128	42	42	NM	127	121	141	285	276	21	22	56	60
20/3	JM	49	48	177	NM	NM	53	175	NM	127	155	344	351	43	NM	113	NM
20/3	NJF	39	39	153	NM	46	46	NM	NM	92	133	285	265	20	NM	60	50
21/3	KMM	45	45	185	NM	50	51	152	NM	114	150	322	335	34	35	122	130

This particular collar will enable us to confirm this hypothesis. The last collar was fitted on to a young 23 year old male which had both tusks intact at Kamboyo (KMM).

Physical Characteristics of Darted Elephants

Physical measurements of the characteristics of the front leg, hind leg, tail, ears, back, shoulders, and tusks was undertaken. Table 2 below summarizes the physical characteristics of the collared elephants. The average spoor diameter for the females and males were 38.33 ±0.33 cm (n=3) for the left leg and 38.0 ±0.58cm (n=3) for the right leg and 47.18 ±2.15 cm (n=6) for the left leg and 47.0 ± 2.21cm for the right leg respectively.

The sampled animals had a tail length of 122.17 ±6.85cm (n=6) for the males and 111.3±9.67cm (n=3) for the females. This compares closely with the universal measurements with an average tail length of 1-1.5m long and an average weight of 10-kgs and sports a 3-foot (1 m) bunch of wiry hair at the end. This is a perfect tool for swatting flies (Holladay, 2008). Males had a mean task length of 106.4±11.48cm (n=6) for the left tusk and 35.25±4.52cm (n=4) for the lip circumference while the females had a left lip circumference of 20.5 ±0.5cm (n=2), right lip circumference of 22.0±0cm (n=2) and a left tusk length of 65.3±7.42cm (n=3).

5.0. Discussion

Collaring Operation Sites

The initial plan was to deploy ten collars to ten individual elephants, four and six in Tsavo East and West National Park respectively. However, only nine collars (four and five in Tsavo East and Tsavo West) were deployed as one of the collars malfunctioned. The whole elephant collaring operation was planned to take five days (19-23 March 2012), with an average of two elephants being collared each day. The operation however took two days and a few hours in the morning of the third day. This was because the only operational KWS helicopter was required to assist put off fires that broke out in Mount Kenya and Abardares forests during the our operation in the Tsavos. As a result, we had to re-organize for a crush programme to ensure we deploy all the nine collars before the helicopter went to the mountain areas to assist the team putting off fires at the Mountain area.

During the operation, males were collared in Emusaya, Ndia Ndaaza and Ithumba areas of Tsavo East National Park respectively. Males in Emusaya and Ndia Ndaaza area were collared as was proposed.

At Ithumba, the initial plan was to collar a female, but the team ended up collaring a male. This was because it was difficult to sex sighted individuals from a helicopter due to thick vegetation (woodland). Also, a female that was proposed to be collared in Sangayaya area of Tsavo East National Park ended up being collared in Thabanguji area. This was because the whole collaring exercise was done in a rush.

From the planning meetings it was agreed that two elephants were to be collared in a day but the team ended up collaring four elephants per day, this resulted to the team missing a female in sangayaya and did it in Thabanguji. In Tsavo west national park a male was collared in Kasigau ranch which was next to Kasigau gate of Tsavo west national park. Elephants were not found in Kasigau gate area of Tsavo west as was proposed and the team decided to select the one who was in the ranch but close to Tsavo west. Males in the jipe and Kamboyo area were collared as was proposed the same to the females in Njukini and Maktau area. In chala area a female elephant was not collared because the collar failed to link to the satellite and it was sent back to Kenya wildlife service Tsavo east research centre to be monitored. During collaring planning meetings a 50:50 males to female ratios were agreed upon, which meant that two males and two females were to be collared in Tsavo east while three females and three males were to be collared in Tsavo west. During the collaring three males and one female was collared in Tsavo east for reasons explained above (Ithumba case) while in Tsavo west three males and two females were collared because one collar which was meant for a female in Chala did not work. In total six males and three females were collared in March 2012 in Tsavo conservation area.

Because of the crush programme, the following was observed:

i. All the drivers had to drive with a very high speed of 85-125km/hr so as to reach or connect from one location to the other, lucky enough no accident occurred during the process but everybody was red in dust after the end of everyday and very exhausted.
ii. The vehicle got a lot of tyre punctures and they had to go to the garage after the end of the operation.
iii. One media vehicle got lost and a KWS vehicle had to look for it and in two occasions the same vehicle veered off the road and almost rolled because the drivers was trying to catch up with the speed at which the KWS vehicles were moving with.
iv. Following the exhaustion, coupled with thick habitat vegetation in Ithumba area, wrong judgment was done when collaring Ithumba elephant. In Ithumba area a female elephant was the target but we ended up collaring a young male elephant of about 17 years of age. Following that the Ithumba male will be monitored for six months thereafter the male will be darted, collar removed and deployed to the female. The veterinary gave a period of six months as the required time for the animal to heal and recover from the capture stress.
v. The rushed programme ended up in collaring Sangayaya female in a different location. The plan was to collar a female in Sangayaya but we ended up collaring the female in Thabanguji. The biologist noted with concern that the Sangayaya-Thabanguji female collar was a little bit tight on the neck. The female will be monitored to ascertain the status of the collar, with the aim of rectifying it during the deployment of the collar that malfunctioned.

Other observations made during the collaring operation were that the Ndia Ndaza male elephant had arrow heads on its body. The arrow was removed by the veterinaries and the wounds cleaned. This was an indication that there was insecurity in Ndia Ndaza areas of Tsavo East National Park. In addition there were lots of livestock and bomas in the same area. The park management needs to deploy security teams in the area to control livestock incursion and reduce insecurity to elephants.

During the operation three elephants were collared outside the protected area and these were the Kasigau male, Jipe male and Njukini female. Three males which were collared had only one left tusk. Those males were in Emusaya, Ithumba and Jipe respectively.

Elephant Age, Spoor Diameters and Length of the Legs

The age of the collared elephants were 26.67 ± 2.69 years (n= 9) with a minimum of 17 years and a maximum of 40 years. This was within planned proposal to deploy the collar on most active age groups of both males (25-30 years) and female (20-25 years) elephants. This was primarily aimed at enhancing coverage of the animals in the study sites. The assumption was that these age groups would generate more information through area coverage, better reflection of the environmental impacts on the animal, more exploratory than the old age and younger age groups. Typically elephants live for 50 to 70 years, but the oldest recorded elephant lived for 82 years. This therefore is inconformity with the selected age groups are optimal growth period for the elephants beyond which minimal growth is observed and the allowance left on the collars is adequately covered to suit the animal without constraining.

Elephant´s foot size is used to judge the overall size of a particular animal, its age, and the number of elephants in a particular group. The circumference of the forefoot is approximately equal to half the shoulder height (Daryl Balfour and Sharna Balfour 1998) while young breeding females are expected to be in relatively bigger herds than the females. Old males are normally solitary and old females stick together with the family.

The average spoor diameter for the females was 38.33 ±0.33 cm (n=3) for the left leg and 38.0 ±0.577cm (n=3) for the right leg while for males were 47.18 ±2.15 cm (n=6) for the left leg and 47.0 ± 2.21cm for the right leg as compared with the general measurement of 40-50cm in length and width and a circumference of about 1.34 meters (Burfour and Burfour, 1998). On variations in spoor diameters, the results indicate insignificant differences between the front legs of the same individual. The same applied for the rear legs. However, some differences between the front and the rear feet, the later being slightly bigger than the former. These variations seem to increase with age so as to support overall increasing weight of the animal. The larger the ears, the more the cooling hence with increasing age/size the need for bigger ears.

Elephant tusks are also an indicator of age and activity of the animals. Both male and female African elephants have tusks, however, only the male species of the Asian elephant has them. Tusks continue growing for most of the elephant life and are therefore an age indicator - much like the elephants feet, the age of the elephant can be estimated by observing their tusks. The size of elephant tusks is an inherited characteristic of an individual animal. Tusks of a baby elephant (calf) measure only about 5cm long. These 'milk tusks' will fall out around their first birthday. Their permanent tusks will then start to protrude beyond their lips at around 2-3 years old and will continue to grow throughout their lives. Tusks grow at about 15-18cm per year. Interestingly, some elephants are born without tusks. This hereditary condition causes huge differences in the musculature and shape of the neck and the head of the elephant. Also, the carriage of the head is different and the bones at the back of the skull are less developed. During this collaring exercise, three (3) males had one left tusk only representing 33% (n=9) of the sampled group as compared to 40-50% of male Asian elephants that are tusk less.

Elephants are either 'left-tusked' or 'right-tusked', just like a human being might be 'left-handed' or 'right-handed'. The favoured tusk is usually shorter than the other due to constant use. Tusks on an individual elephant can differ in shape, length, thickness, and growth direction. Male elephants tend to have heavier, longer and stouter tusks than females do.

Animal Welfare during Collar Deployment

Reducing stress to the darted animals is paramount in operations dealing with wildlife. Animal welfare is a key consideration and to achieve this, extensive planning is done to ensure that all necessary equipment and expertise is employed in the exercise. During the planning a deliberate effort is made to identify required teams and specific roles assigned followed by proper briefing before the exercise. These personnel is usually inclusive of expertise from experienced wildlife biologists and veterinarians, veterinarian technicians, wildlife capture teams, wildlife capture pilots, logistical personnel, wildlife protection rangers and experienced drivers in wildlife conservation. A communication protocol is established and adhered to during the operation to minimize incidents that could potentially threaten the animal's welfare. Darting is usually scheduled for cool hours of the day that is early morning and late evening. The dosage administered to animals is usually determined by the age, sex and the demeanor. The main objective is usually to immobilize and collar the animal under the least stress levels.

Once the target group has been identified by the spotting team, this is communicated to the darting team in the Chopper aircraft where highly experienced wildlife biologists have to positively identify the best candidate taking into consideration the sex, age, body condition, in the case of female candidates' great care is taken to avoid those with obvious signs of pregnancy, matriarchs, those nursing young ones below the age of two years and for males sub adults less than 20 years are not targeted because it is still growing and the neck circumference could be interfered with by the collar.

Once the wildlife biologists are satisfied they have the right candidate, they communicate this to the veterinary and chopper pilot who now coordinate to ensure the individual is darted most preferably on the rump. After successfully darting the candidate, the chopper team carefully monitors the animal and once it is immobilized, the ground team is called in, where the capture team which includes a veterinary officer is the first to arrive to ensure that the animal is lying in the right position before collaring operation is initiated. A lateral recumbent position is preferred and when an animal lies is in sternal position the capture team has to reposition it. Certain risks associated with this operation include heat stress, animal falling on bad terrain or in sitting position or the interference with the breathing if the trunk is twisted or closed up and hence deliberate measures are taken to minimize this. Noise is avoided as this can cause undue stress to the animal. The collars are deployed as quickly as possible so that the immobilized animal can be revived in the shortest time possible. It is important to note that if a darted individual is observed to show stress signs, the veterinary officer in-charge immediately revives the animal and the search for another candidate is initiated. The immobilization drug used also interferes with the animals thermal regulation hence adequate water is poured on the animal especially at the back of the ears where there high blood supply to cool the animal.

In this operation the first four elephants collared on day 1 had an average collar deployment time of 13 minutes, while on the second day there was improved efficiency in the operation reducing the average collar deployment time to 10 minutes. In the end there were no injuries or deaths to elephants or personnel involved in this inherently dangerous exercise. This remarkable efficiency is testimony to the thorough preparations of the last six months and the experience of the technical and support teams involved in the operation.

CONCLUSION

The current challenge of elephant poaching and human-elephant conflicts requires information on use of space by the elephants. This is important to wildlife managers as it provides them with information required to adequately deploy security and problem animal control teams. This information can only be available by deploying satellite-linked GPS collars to the elephants. We report that nine satellite-linked GPS collars were deployed to nine elephants (3 bulls and 1 female in TENP; 3 bulls and 2 females in TWNP) during the operation. During the operation, neither elephants nor humans were injured.

The age of the collared elephants ranged from 17 years to 40 years. This age was almost within the recommended age category (about 20 years) for elephants that can have collars deployed around their neck. The elephants were successfully collared at Ndia-Ndaasa (male=1), Thabanguji (female = 1), Ithumba (male = 1), and Emusaya (male = 1) areas of TENP as well as Kamboyo (male=1), Kasigau (male=1), Jipe (male=1), Maktau (female=1), and Njukini (female=1). Monitoring of these elephants will continue for the next 20 months to know their movement patterns and map their migratory corridors and dispersal areas.

RECOMMENDATIONS

From the operation, the following is recommended:

a) It is important to ensure that the planned darting program is strictly followed while sticking to the agreed upon collaring site, sexes and age class of individuals.
b) Darting around mid-day or early afternoon should be avoided in future.
c) Contingency plans for an alternative helicopter should be organized during the planning stage.
d) There is need to use coordinates of the area, instead of using the general area name. Maps should be provided to all the team leaders so that they understand which areas the operation is to take place.
e) More data (e.g., location of water points, vegetation characteristics or salt licks) should be taken by the search and darting teams.

ANNEXES

Annex 1. Do's and Don'ts during a Collaring Operation

1. Personal Safety

- Ground teams should not move to site of immobilized animal until the ground team leader confirms that the animal is down and immobile.
- Only the capture team, data collection team and collar deployment team should be at site where the immobilized elephant has fallen. The other teams should be about 10 meters from the fallen animal, whereas the security teams should be about 15-20 meters from the fallen animal.
- No person or vehicle should be at the front side of the elephant.
- All persons should be at safe distance in a vehicle before the revival drug is administered.
- Crowding around immobilized elephant is discouraged
- Noise should be avoided at all times.
- In case of a snake bite, as much as possible, try and take note of the size of the snake, its length, its skin colour and the shape of the head. This will help in determining the antidote.
- Every person should be cautious enough during the operation to ensure personal security.
- Do not use strong perfume.
- In case of bees, be calm and calmly lie down.
- Wear boots reaching up to your ankles.

2. Helicopter

- Approach the helicopter from the front not at the rear.
- Once you alight from helicopter, always head towards the front.
- The capture sergeant in the helicopter should be armed and move to ground with his firearm.
- Once the helicopter lands, it should be guarded by two rangers to avoid attack by elephants.

3. Security

- Security teams should guard the site of immobilized elephant facing the source of danger, always facing away from the elephant.

4. Vehicles

- All vehicles should have their doors open once on site where animal has fallen down.

- Park vehicles facing away from the animal
- All vehicles should be parked with a clear escape route.
- All drivers should remain in their vehicles.
- There should be at least a vehicle with a working winch.

5. Media
    - Do not insist on remaining on the ground to take photographs when the animal is standing up after being revived.

6. Drugs
    - Do not handle the immobilization or revival drugs, only a qualified veterinary doctor should handle them.

7. Animal Welfare
    - Adequate water for cooling animal must be availed.

8. Collar Team
    - All operational equipment and materials should be collected from site after the operation is complete.

Annex 2. Collaring Teams

1. Spotting Team

    (i). Pilot (TENP and TWNP
    (ii). B. Ngoru – TENP/C. Khayale – TWNP

2. Chopper (Darting) Team

    (i) Helicopter Pilot
    (ii) Dr. J. Poghon (darting)
    (iii) S. Njumbi (identification of elephant to be collared)
    (iv) D. Kimutai (identification of elephant to be collared)
    (v) Camera Man
    (vi) Capture Sergeant Lelimo

3. Ground teams

    a) Collar Deployment

    (i). P. Mulandi (collar fitting)

(ii). Dr Poghon (animal welfare and reviving)
(iii). Capture Sergeant Lelimo (collar fitting)
(iv). E. Bitok (neck measurement and collar fitting)
(v). D. Muteti/Nyagah (neck measurement+collar fitting)
(vi). S.Njumbi (collar fitting and timing)
(vii). Dr. Ndeereh-HQs Vet (animal welfare)

b) Security

(i). Company Commander TE/TW
(ii). NCO
(iii). 5 Rangers
(iv). Driver

c) Data collection (2 vehicles)

a. E. Bitok (oversight + tagging)
b. D. Muteti/Nyagah (measurements + tagging)
c. H. Syengo/S. Muli (watering animal)
d. A. Mwazo and M. Nzisa (Timing and Measurements)
e. Kabete/Munyoki (data recording)
f. Ranger Mutisya (TWNP)/Ranger Gitau (TENP), I. Peroh – Painting

4. Transport Management

(i) N. Wekesa
(ii) P. Namunyu,
(iii) F. Ojwang
(iv) Mechanic
(v) Telecom technician

5. Radio Communication

(i). F. Ojwang

6. Communications/Media

(i) E.Wamba (IFAW)
(ii) O. Mulei-KWS
(iii) Journalists (X 4)

7. Logistics and Administration

(i). G. Kariuki (TENP)
(ii). G. Muraya (TWNP)

(iii). B. Awayo (IFAW)
(iv). I. Peroh (IFAW)

8. First Aid

(i) Rgr. Tirop TE and Cpl Azere TW

9. Aircraft technician and attendants

Annex 3. Chain Of Command

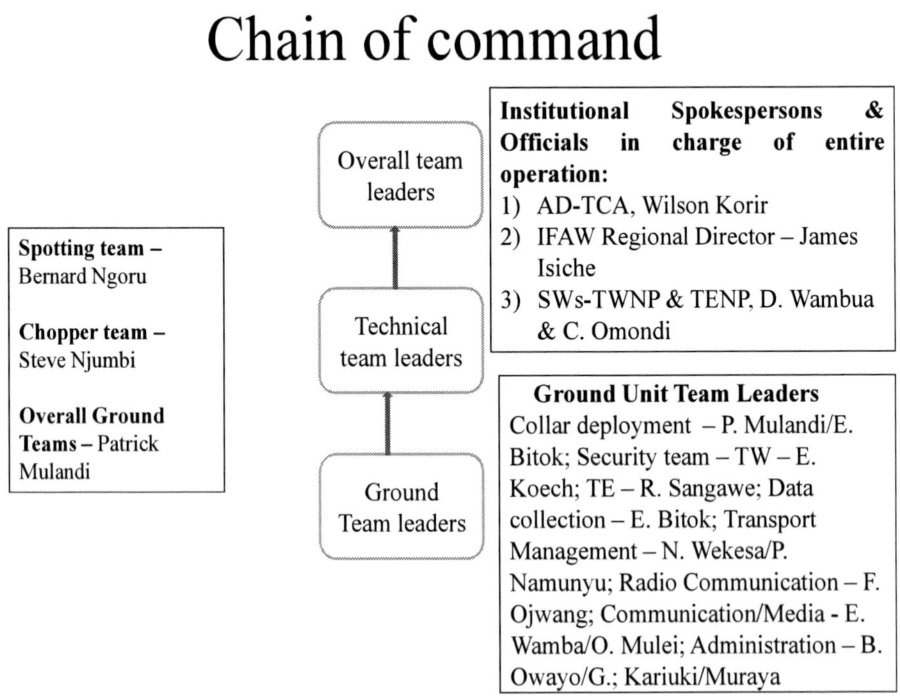

Annex 4. Order of Approach by Ground Team

1. The vet doctor to dart the selected animal. Chopper Lands when animal is down. Vet to attend before ground team arrive
2. The Capture Warden and Ground Vet to secure the elephant
3. The collar deployment team
4. The Security team to secure the area
5. The second biologists to collect all biophysical data from darted elephant; The vet doctor to administer antidote to the elephant once all is complete
6. Rest of team in position away from the elephant
7. Once Collar is fitted and all measurements are taken. All to clear scene except the Vet

Annex 5. Elephant Collaring Tools/Equipments

1. GPS/satellite collar
2. Wire to pull collar
3. Bolts and Nuts
4. Ratchet Spanner
5. Tape measure
6. Paint
7. Brushes for painting
8. A sharp knife/panga/scissor
9. Plates
10. Data Sheets
11. Tags
12. Water for cooling
13. Power saw
14. Stop Watch

Annex 5. Collar Deployment Techniques

- Secure the elephant – ensure the elephant is lying in the correct position.
- Measure the neck circumference – give an allowance of 6 inches
- Measure the collar and cut as appropriate
- Pull the collar underneath the neck using a hooked wire
- Attach the Lead Weight using the plates, nuts and bolts
- Ensure the counter weight is well balanced with the upper unit.

Annex 6. Data Collected

- Shoulder height
- Back length
- Tusk Measurements
- Spoor Diameter
- Tail length
- Age and Sex
- Elephant ID - painting
- Collar frequency and ID
- Animal behavior before and after darting and revival
- Blood and parasite samples
- GPS location
- Position of dart and range
- Type of dart Gun used
- Timings

Annex 7. List Of Participants

The following personnel participated in the operation:

1. Kenya Wildlife Service(KWS)

   a) Wilson Korir, Assistant Director Tsavo Conservation Area- *Institutional Spokespersons and* overall in-charge of the operation
   b) Dominic Wambua: Senior Warden-Tsavo east, in-charge of the operation
   c) Collins Omondi: Senior Warden-Tsavo west, in-charge of the operation
   d) Bernard Ngoru: Senior Research Scientist - TCA, in-charge of operation's logistics and Spotting
   e) Elphas Bitok: Research Scientist-Research scientist, Southern conservation area in charge elephant collaring, oversight + tagging
   f) D. Kimutai, Research Scientist, elephant programme, identification of elephant to be collared
   g) Jeramiah Poghon: Veterinary Officer, animal immobilization and darting
   h) Ndereeh: Senior veterinary Officer- Head vet and monitoring condition of immobilized elephants, animal welfare
   i) Kariuki, Lab technician- Veterinary Sample collection
   j) P. Mulandi, Warden capture- in charge of capture team and collar fitting
   k) Captain Kiroken – Helicopter Pilot
   l) Captain Shamalla and Mwangi – Fixed Wing Pilot
   m) D. Muteti/Nyagah (measurements + tagging
   n) H. Syengo/S. Matheka (watering animal)
   o) M. Nzisa (Timing and Measurements)
   p) Kabete/Munyoki (data recording)
   q) Ranger Mutisya/Ranger Gitau– Painting
   r) Koech and Olemwaniki: Security officers, coordinating team security
   s) Rangers from KWS Tsavo East: Security and Safety, backup personnel
   t) Daniel Lelimo: Capture Unit , secure elephant and assist collaring
   u) N. Wekesa and P. Namunyu, workshop Manager- in charge of transport Management
   v) F. Ojwang, Senior telecom officer- In charge of radio communication
   w) Obedi Mulei, in charge of communications/Media
   x) G. Kariuki and G. Muraya, Park administrators- in charge *of logistics and Administration*
   y) Rgr. Tirop TE and Cpl Azere TW- in charge of *First Aid*
   z) Aircraft technician and attendants in Tsavo

2. International Fund for Animal Welfare(IFAW)

   a) Steve Njumbi Operational coordination and Collaring
   b) James Isiche: Overseeing operations
   c) Alex: data collection
   d) Beatrice Awayo: Logistics

e) Isaac Peroh – Procurement
f) Elizabeth Wamba- Media andcommunications

Annex 6. Immobilization, Animal and Personnel Safety

1. Method of Immobilization

    - Chemical Restraint

        - Use of drugs that anaesthetize, sedate, tranquilize or hypnotize an animal.

    - Drugs will be delivered by darting from a Chopper
    - Choice of drugs:

        - Immobiliser:Etorphine
        - Revival: Diprenorphine

    - Dosage:depends on species, age, sex and demeanor.

2. Safety Of The Animal

    - Immobilization will be scheduled for cool hours of the day, early morning and evening hours.
    - Objective is to quickly immobilize and collar the candidate under the least stress levels.
    - Risks associated with the animal are heat stress, falling on a bad terrain i.e. rocks, sloppy ground or in sitting position.
    - Adequate water for cooling must be availed

3. Personnel Safety and Security

    - The vet will always be the first to the animal and last to leave.
    - Front side of the elephant should be clear of personnel or vehicles.
    - Crowding around the immobilized elephant is discouraged.
    - In case of any danger to the animal, immediate revival will be initiated and all personnel will have to vacate.
    - Noise should be avoided at all times, semi-anaesthetized individuals can easily be aroused by noise, irritable and can be unpredictable especially the females.
    - Security officers should be on alert throughout the operation.
    - Drugs used in immobilization are highly toxic to humans. Only to be handled by Trained personnel.

Annex 7. Collaring Operation Programme

Day/date	Time	Activity	Responsibility
Monday- Friday 19-23 Mar 2012	5:30-6:00a.m.	Breakfast, preparation of aircrafts for takeoff at 6:00a.m. Ground team departs before 6:00a.m	All participants
	6:00a.m.	The chopper and light aircraft take off	Pilots
	6:00a.m. -11:00a.m.	Search for elephant groups, immobilize elephants, deploy collar, collect biophysical data, revive elephant	All participants
	11:00a.m.-3:00p.m.	Lunch, debriefs if need be, movement to afternoon sites if need be	All participants
	3:00p.m.-5.00p.m	Search for elephant groups, immobilize elephants, deploy collar, collect biophysical data, and revive elephants	All participants
	5:00-6:30p.m.	Debrief, planning for the following day's operation	Technical team leaders

Annex 8a. Table Showing the Locations Where the Elephants Were Collared, the Name of Veterinary That Tranquilized the Elephants, the Type of the Gun He Used and the Distance of the Elephant from the Chopper When It Was Darted

Vet name	Dart gun type	Range	Date	Elephant name	GPS Position	
					Latitude	Longitude
Poghon	Dan-Inject	30m	19-03-12	Emusaya male	37m 0496792	UTM 9690092
Poghon	Dan-Inject	40m	19-03-12	Ndiandaza male	37m 0478418	UTM 9735845
Poghon	Dan-Inject	30m	19-03-12	Ithumba male	37m 0432815	UTM 9749864
Poghon	Dan-Inject	30m	19-03-12	Sangayaya female	37m 0459456	UTM 9677041
Poghon	Dan-Inject	20m	20-03-12	Kasigau male	37m 0449691	UTM 9573577
Poghon	Dan-Inject	25m	20-03-12	Maktau female	37m 0397754	UTM 9627665
Poghon	Dan-Inject	20m	20-03-12	Jipe male	37m 0364735	UTM 9602658
Poghon	Dan-Inject	30m	20-03-12	Njukini female	37m 0356997	UTM 9653272
Poghon	Dan-Inject	20m	21-03-12	Kamboyo male	37m 0402814	UTM 9692865

Annex 8b. Table Showing the Dosages which Were Used to Tranquilize the Elephants, the Area Where the Dart Had Hit and the Position the Elephants Fell with When They Went Down

Elephant name	Position of the dart	Time darted	Dart dosage (etorphine)	Time down	Position elephant fell
Emusaya male	Rump	8:15 AM	18 mg	8:25 AM	Right side
Ndiandaza male	Rump	10:26 AM	17 mg	10:35 AM	Left side
Ithumba male	Rump	2:25 PM	17 mg	2:27 PM	Left side
Sangayaya female	Pelvis	4:37 PM	17 mg	4:47 PM	Left side
Elephant name	Position of the dart	Time darted	Dart dosage (etorphine)	Time down	Position elephant fell
Kasigau male	Rump	7:59 AM	18 mg	8:09 AM	Left side
Maktau female	Rump	10:28 AM	17mg	10:36 AM	Left side
Jipe male	Rump	12:27 PM	18 mg	12:37 PM	Right side
Njukini female	Rump	3:27 PM	18 mg	3:31 PM	Sternal then pushed and lay on right side
Kamboyo male	Rump	8:41AM	18 mg	8:48 AM	Right side

Annex 8c. Table Showing the Revival Dosages Administered to the Elephants and the Time the Elephants Took to Wake Up after the Revival Drugs Were Administered

Elephant name	Time elephant took to wake up when revival was injected	Revival dosage (diprenorphine)	Time up	Other drugs
Emusaya male	2 min 10 sec	54mg	8:44 AM	1000 iu hyalase
Ndiandaza male	1 min 50 sec	51mg	10:58 AM	1000 iu hyalase
Ithumba male	Not recorded	51mg	2:49 PM	1000 iu hyalase
Sangayaya female	1min 53sec	51mg	5:00 PM	1000 iu hyalase
Kasigau male	2min 26sec	54 mg	8:25 AM	1000 iu hyalase
Maktau female	1min 50sec	51 mg	10:49 AM	1000 iu hyalase
Jipe male	3min 23sec	54 mg	12:54 PM	1000 iu hyalase
Njukini female	2min 1sec	54 mg	3:45 PM	1000 iu hyalase
Kamboyo male	2min 6sec	54 mg	9:02 AM	1000 iu hyalase

Annex 8d. Table Showing Other Samples which Were Taken During Collaring

Other biophysical information (measurements in centimeters)				
Elephant name	Tusk length R	Blood samples	Tick samples	Other samples
Emusaya male	Not measured	EDTA Blood	No ticks	Tail hairs
Ndiandaza male	148	EDTA Blood	Ticks	Tail hairs
Ithumba male	No tusk	EDTA Blood	No ticks	Tail hairs
Sangayaya female	80	EDTA Blood	Ticks	Tail hairs
Kasigau male	105	EDTA Blood	No ticks	Tail hairs
Maktau female	60	EDTA Blood	Ticks	Tail hairs
Jipe male	Not measured	EDTA Blood	No ticks	Tail hairs
Njukini female	50	EDTA Blood	Ticks	Tail hairs
Kamboyo male	130	EDTA Blood	Ticks	Tail hairs

Annex 8e. Table Showing the Time when the Collaring Team Started Deploying Collars on the Elephants, Time Taken to Do the Collaring on Different Elephants and the Vhf Frequencies of the Different Collars

Collars and other collaring information			
Elephant name	Time when collar was started to be deployed	Time taken to do the collaring	Collar frequency
Emusaya male	8:29 AM	13min	150.590
Ndiandaza male	10:36 AM	25min	150.110
Ithumba male	2:35 PM	14min	150.130
Sangayaya female	4:46 PM	11min	150.650
Kasigau male	8:11 AM	9min 20sec	150.930
Maktau female	10:38 AM	9min 9sec	150.890
Jipe male	12:39 PM	12min 38sec	150.040
Njukini female	3:33 PM	10min	150.910
Kamboyo male	8:50 AM	9min 29sec	150.060

Annex 8f. Table Showing Collar Id's, the Type of the Collars Used, the Elephant Id's and Ages of Different Elephants which Were Collared

Collars information				
Elephant name	Collar ID	Collar type	Elephant ID	Age of the Elephants (years)
Emusaya male	IFAW 14	Satellite –GPS/VHF	EM	30
Ndiandaza male	IFAW 12	Satellite –GPS/VHF	NM	34
Ithumba male	IFAW 13	Satellite –GPS/VHF	IM	17
Sangayaya female	IFAW 8	Satellite –GPS/VHF	SGF	23
Kasigau male	IFAW 17	Satellite –GPS/VHF	KM	40
Maktau female	IFAW 15	Satellite –GPS/VHF	MKF	21
Jipe male	IFAW 9	Satellite –GPS/VHF	JM	37
Njukini female	IFAW 16	Satellite –GPS/VHF	NJF	23
Kamboyo male	IFAW 10	Satellite –GPS/VHF	KMM	23

REFERENCES

Blanc, J.J., Thouless,C.R., Hart, J.A., Dublin, H.T., Douglas-Hamiliton,I., Craig, G.C.and Barnes, R.F.W. (2007) African elephant status report 2007: An update from the African Elephant Database. Occasional paper of the IUCN Species Survival Commission No.33. Gland, Switzerland and Cambridge, UK: IUCN.

Cobbs, S. (1976) The Distribution and Abundance of the Large Herbivore Community of Tsavo National Park. D.Phil. Thesis Univ. of Oxford.

Balfour, D. and Balfour, S. (1998) African Elephants: A Celebration of Majesty. Publisher: Abbeville Press.

Douglas-Hamilton, I. Gachago, S., Litoroh, M., Mirangi, J. (1994) Tsavo Elephant Count 1994. Kenya Wildlife Service.

GOK, (2010). 2009 Population and Housing Census Results. Kenya Bureau of Statistics.

Inamdar, A. (1996) The Ecological Consequences Of Elephant Depletion. Unpublished PhD thesis, University of Cambridge.

Holladay A. (2008), USATODAY.com, a division of Gannett Co. Inc.

IUCN (1978) Results of an Aerial Census of the Tsavo National Park and Ecosystem. Typescript.

Kahumbu, P., Omondi, P., Douglas-Hamilton, I., King, J. (1999) Total Aerial Count of Elephants in Tsavo National Park and Adjacent Areas. Kenya Wildlife Service.

Kangwana, K. (1996) Studying elephants. AWF technical workbook, Nairobi, Kenya. ISBN: 9966-9915-0-6.

Kenya Wildlife Service (2012) Elephant Mortality database. KWS, Nairobi, Kenya.

Kenya Wildlife Service (2008). Environmental Impact Assessment report for the Jipe-Rombo Electric fence. Unpublished report.

Leuthold, W. (1976) Age structure of elephants inTsavo National Park, Kenya. *Journal of Applied Ecology*, 13, 435-444.

Leuthold, W. and Sale, J.B. (1973). Moments and Patterns of Habitat Utilization of Elephants in Tsavo National Parks, Kenya. *East African Journal*, 11, 369-384

Ngene S., Njumb S., and Omondi P. (2011). Total aerial census of elephants and other large mammals in the Tsavo-Mkomazi ecosystem. Kenya Wildlife Service.

Olindo P, Douglas-Hamilton I, Hamilton P. (1988) Tsavo elephant count 1988. Nairobi: Wildlife Conservation and Management Department and Ministry of Tourism, Kenya.

Omondi, P.O, Bitok, E., and King, J. (2002) Total aerial count of elephants and buffalo in the Tsavo/Mkomazi ecosystem Nairobi: Kenya Wildlife Service.

Omondi, P. and Bitok, E. (2005) Total Aerial Count of Elephants, Buffalo and other species in the Tsavo/Mkomazi Ecosystem. KWS/ MIKE Programme Report.

Omondi, P., Bitok E.K., Mukeka, J., Mayienda, R.M. and Litoroh, M. (2008) Total aerial count of elephants and other large mammal species of Tsavo/Mkomazi ecosystem. Kenya Wildlife Service.

WCMD (1976) Report of Aerial Counts of Tsavo. Typescript.

Wijgaarden W. (1985) Elephants – trees- grazers: Relationship between climate, soil, vegetation and large herbivores in a semi-arid ecosystem (Tsavo, Kenya). ITC Publication #4.

In: Recent Advances in Satellite Research and Development ISBN: 978-1-62417-443-8
Editors: Samuel Gardiner and Kenneth P. Olsen

Chapter 2

ENHANCING RESILIENCY OF BROADBAND SATELLITE COMMUNICATION EMERGENCY NETWORKS THROUGH PROPAGATION IMPAIRMENT MITIGATION TECHNIQUES

C. I. Kourogiorgas**, V. K. Sakarellos, S. A. Kanellopoulos and A. D. Panagopoulos
School of Electrical and Computer Engineering
National Technical University of Athens
Iroon Polytechniou, Zografou, Athens, Greece

ABSTRACT

In this chapter the main issues raised from the employment of telecommunications satellites at emergency communications operating at high frequencies are described. First, the need of telecommunication systems in emergency situations is analyzed showing that these are mandatory for the disaster relief. More specifically, the chapter is focused on the presentation of the various telecommunication services which must be supported, as well as for the backhauling of these networks, by satellite systems. Since, broadband satellite communication systems operate mainly above 10GHz, the propagation impairments at these frequencies are described. They depend mainly on precipitation, atmospheric gases, clouds and tropospheric turbulence. Apart from models for the prediction of the first-order statistical characteristics, models for the generation of time series are also presented. Great attention is paid on rain attenuation, since it is the dominant fading mechanism for the satellite communication systems. Moreover, the most important propagation impairment mitigation techniques that have been proposed in the literature are presented in detail. Finally, general concepts for future research are given and some conclusions are derived.

* E-mails: {harkour, bsakarel, skanello}mail.ntua.gr, thpanag@ece.ntua.gr

1. INTRODUCTION

In the last years, many disasters and emergency situations have occurred all around the globe. The disastrous Earthquakes in Greece and Turkey (1999), the Katrina hurricane in USA (2005), the two tsunamis in Indonesia and Japan (2004, 2011), the forest fires in Greece (2007) are a few among others catastrophic situations in the last fifteen years. According to ETSI EMTEL group [1], as this is stated in [2], emergency is "*an urgent need for assistance or relief*" and emergency situation may include everyday incidents like traffic accidents and major incidents and disasters like forest fires and earthquakes.

In [2], emergencies are categorized as (a) daily emergencies and (b) disaster emergencies which are "*a serious disruption of the functioning of society, posing a significant, widespread threat to human life, health, property or the environment, whether caused by accident, nature or human activity, and whether developing suddenly or as the result of complex, long-term processes*".

In [3], two more categories of emergency situations are defined (similar with those in [2]) without considering the cause of the emergency situation: a) "catastrophe with limited impact" and b) "catastrophe beyond response means". A situation can be included in one or the other category whether the regional response forces are able to deal with the situation or not and depends on the nature of the event, the number of threaten citizens etc.

In emergency situations, the telecommunications are of utmost importance since they offer the communication between the rescue teams and their coordinators, as well as the communication between the victims with their relatives and the rescue teams. However, in many situations, the terrestrial telecommunications infrastructures are destroyed due to the disaster or they do not exist, for example in forest fires and earthquakes. Therefore, satellite communications seem to be the most reliable way for the communication during an emergency, as they do not get damaged by the disaster.

The role of a satellite during emergency communications will be analyzed in a next section of the chapter. Telecommunication services for emergency situations have been identified by the Radiocommunication sector of International Telecommunication Union (ITU-R) in [4] and these are presented in Table 1.

The services are different for the various phases of emergency according to ITU-R. As it can be observed from this table emergency telecommunications are really significant in order to deal with catastrophic situations.

In disaster management, at least three phases have been identified, according to ETSI standard [2]:

1. Preparedness, i.e. the telecommunications infrastructure used in an emergency must be able to operate and Earth observation for hazard detection.
2. Emergency response which can be defined as "the reaction of Society to a disaster" and it contains all the period when lives can still be saved. In [3], the emergency response starts after the disaster and may last hours or even days depending on the disaster.
3. Return to normal situation which is actually a post-disaster phase, often called "Recovery".

EMERGENCY RESPONSE

Emergency response as defined in the previous section is the response of the Society to a disaster. As explained in detail in [3], the specific organization that is mainly leading the emergency response (e.g. civil protection authority) should efficiently control the telecommunications architecture since the efficiency of the emergency response depends on the training of the forces that actually take action during this phase in order to deal with the results of the disaster. A second reason is that the process of the emergency response is governed by national or regional laws.

Figure 1 shows the state diagram of emergency response as this is recommended by ETSI standard showing the need of telecommunications in emergency situation. In this figure, four states are observed the "*vigilance*", "*anomaly assessment*", "*operations*" and "*alert*".

In the state of vigilance there are two main operations: the detection and location of anomaly and threat and the communication between authorities. The detection of hazards may occur e.g. either from a call of a citizen or a systematic watch from professionals or via sensors. Especially, sensors are useful in cases where people cannot go. The location of the disaster can be pointed through the GNSS and GPS/Galileo/Glonass constellations. The vigilance system must interact with the anomaly assessment state, as the detection and the location of a disaster and possible threat must be assessed by the authorities in order to proceed to the state of the alert.

In the "*alert*" state, there are two scenarios: the alert to authorities and the alert to citizens. The first scenario may be extended to some professionals, which must be alerted at the very beginning like the responsible of industrial plants. The alert to citizens is a responsibility of the authorities. In [2], it is noted that the alert state makes sense in case that the delay between the detection of the hazard is such that people can escape.

In the operations, the involved parties include the administration authorities, which are fixed and responsible for the operations and a fixed temporary task force for coordination.

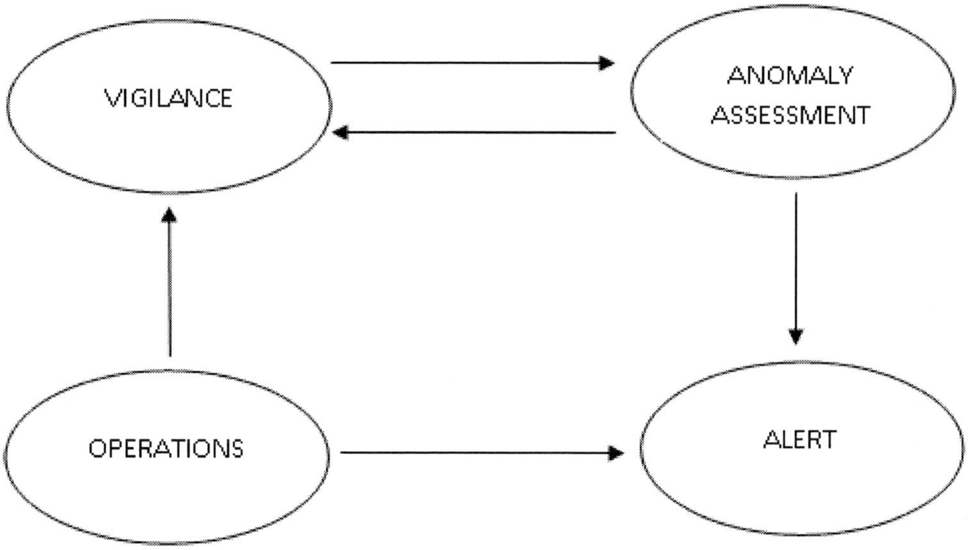

Figure 1. State diagram of emergency response derived from [2].

Table 1. Services in Emergency Radiocommunications

Disaster Phases	Major Radiocommunication Services Involved	Major Tasks of Radiocommunication Services
Prediction and Detection	Meteorological services (meteorological aids and meteorological-satellite service)	Weather and climate prediction. Detection and tracking of earthquakes, tsunamis, hurricanes, typhoons, forest fires, oil leaks, etc. Providing Warning information
	Earth-exploration-satellite service	
Alerting	Amateur services	Receiving and distributing alert messages
	Broadcasting services terrestrial and satellite (radio, television, etc.)	Dissemination alert messages and advice to large section of the public
	Fixed services terrestrial and satellite	Delivering alert messages and instructions to telecommunication centres for further dissemination to public
	Mobile Services (land, satellite, maritime services, etc.)	Distributing alert messages and advice to individuals
Relief	Amateur services	Assisting in organizing relief operation in areas (especially when other services are still not operational)
	Broadcasting services terrestrial and satellite (radio, television, etc.)	Coordination of relief activities by disseminating information from relief planning teams to population
	Earth exploration-satellite service	Assessment of damage and providing information for planning relief activities
	Fixed services terrestrial and satellite	Exchange of information between different teams/groups for planning and coordination relief activities
	Mobile services (land, satellite, maritime services, etc.)	Exchange of information between individuals and/or groups of people involved in relief activities

The fixed temporary task force is responsible for the coordination of the operation and must be far from the field of the disaster and in direct contact with the authorities. Moreover, representatives from the various authorities should also participate.

In some cases, a transportable field task force may be needed closer to the field of the emergency situation for coordination purposes which also communicate with the coordination task force. Another party involved in the emergency is the emergency control centers which are responsible for the coordination of rescue means of each discipline. As in the case of task force transportable emergency controls can be placed close to the field. Moreover, the rescue teams and the citizens are involved in the emergency response. In Figure 2, the communication flow during the operations state in case of emergency is shown and has been derived from [2].

Once the authorities decide to act, the rescue teams are alerted and they are coordinated by the task forces and emergency control centers. The communication, as shown in Figure 2 is between the rescue teams and the near-field coordinators and between the near-field and the far-field coordinators. In [2], it is stated that it is worthy creating a cell around the field of operations of the rescue team in order to communicate with terrestrial mobile radio, since the communication devices can be transported easily on the field. Therefore, many communication technologies such as GSM, Wi-Fi, WiMAX, etc. can be recreated on the field.

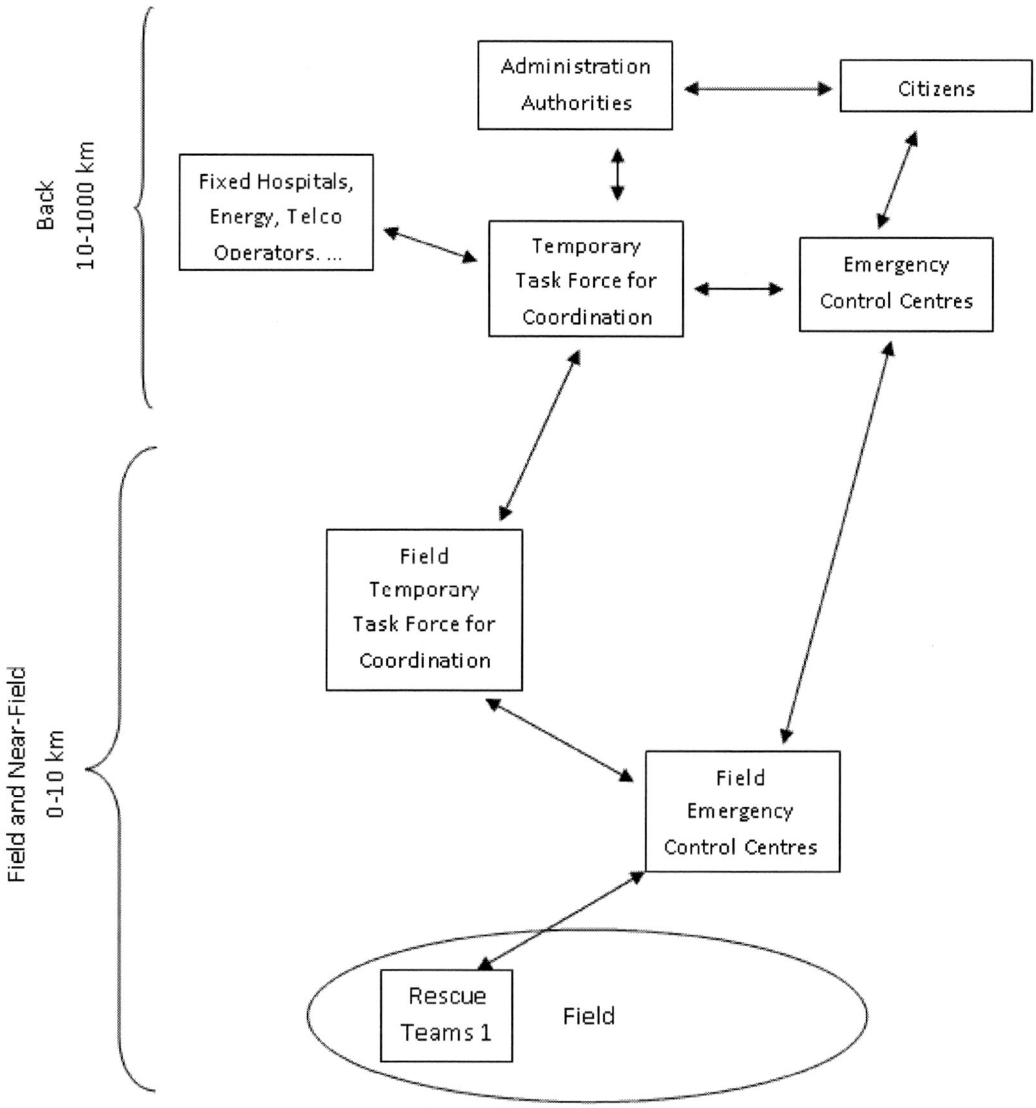

Figure 2. Communications flow during an emergency situation derived from [2].

Some telecommunication services that are required for an emergency were shown in Table 1. However, in operations the nature of the messages can be [2]:

- Basic signals such as emergency calls and broadcast of alerts
- Speech
- Video teleconferencing
- Images
- Data, which are very useful for providing information

After the crisis passes, the situation should return to normal conditions. In the return to normal situation period, the rescue teams must still be coordinated and communicate with the coordinators and also the same services, as before the crisis, must be offered.

As it has been shown the telecommunications are crucial for the emergency situations. Two types of systems have been identified in [3]: the global/continental telecommunication systems which will also be able to be used during an emergency and the small-scale infrastructures which are transported to the location of the emergency and are deployed in order to support the communications.

3. The Role of Satellite in Emergency Communications

3.1. General Concept

It is obvious from the previous section and the introductory part that telecommunications play a substantial role in disaster relief and emergency phases. In this section, the role of satellite is analyzed and highlighted in emergency communications. The satellites can be classified into three categories according to their purpose as *observation*, *navigation* and *telecommunications* satellites. All these three categories can be used in emergency communications.

Earth observation satellites can be used in emergency situations, mainly for the detection of hazards. The role of such satellites has been analyzed in [2] and the ITU-R Recommendation RS 1859 [5]. ITU and ETSI have listed the actions of the observation satellites:

- Identifying areas at risk
- Detecting and tracking earthquakes, tsunamis, hurricanes, forest fires, oil leaks, etc.
- Forecasting weather and predicting climate change
- Forecasting events like El Nino
- Assessing the damage caused by such disasters
- Providing information for planning relief operations
- Monitoring recovery from a disaster

Navigation satellites such as the GPS/Galileo/Glonass satellites can be used for pointing the locations of the disasters and for navigation purposes which are very useful for the rescue teams. Moreover, they can be used for monitoring potential threats. Finally, the telecommunications satellites are useful since they are the most suitable for broadcasting messages to fixed or mobile terminals and for the communications between the terminals on the field of the disaster with the infrastructures at the far-field. Furthermore, since terrestrial communications are usually destroyed, satellites can be used for the backhauling of the communications inside the field and for the communications between the coordination teams and the rescue teams. On demand satellite communication networks can operate immediately and replace successfully the terrestrial telecommunication infrastructure.

The greatest advantage of satellites is that they are not destroyed from a disaster. However, the gateways or the network control centers may be broken down due to a disaster. Placing more gateways and being served from another network control center the problem could be overcome. Moreover, due to the wide coverage of geostationary satellites, they are the most capable to serve the communication needs of the operational field (near- and far-field). Furthermore, since satellites are using spot beams, the capacity can be easily controlled and the need for high bit rates can be fulfilled. In [3], the services of telecommunications satellites are also outlined:

- Providing terminals with mobile and long haul capabilities for the first moments of the disaster
- Backhauling mb/s of data over large areas
- Providing temporary links for the disconnected networks
- Broadcasting of data to large areas

A recommended use of telecommunications satellites for the coordination of rescue teams has been reported in [2].

As it was explained in the Section 2 in operations state it is useful to establish a transported network such as GSM, Wi-Fi, WiMAX, etc. inside the field of the disaster for the communications between the rescue teams and the citizens. However, the communication between the near-filed and far-field operators will be realized through a satellite. Furthermore, since the terrestrial infrastructures may have been totally degraded and maybe destroyed, the satellite communications infrastructure may be used for the communications when the situation has returned back to normal.

ITU has already published the recommendation ITU-R S. 1001 for frequency allocation for fixed satellite systems for emergency communications and it is reproduced here in Table 2.

Table 2. Frequency ranges for fixed-satellite service in emergency situations

Frequency (GHz)	Frequency Allocation		
	Region 1	Region 2	Region 3
3.4-4.2	(space-to-Earth)	(space-to-Earth)	(space-to-Earth)
5.725-5.85	(Earth-to-space)		
5.85-6.7	(Earth-to-space)	(Earth-to-space)	(Earth-to-space)
10.95-11.2	(space-to-Earth)	(space-to-Earth)	(space-to-Earth)
11.45-11.7	(space-to-Earth)	(space-to-Earth)	(space-to-Earth)
11.7-12.2		(space-to-Earth)	
12.2-12.5			(space-to-Earth)
12.5-12.75	(space-to-Earth)		(space-to-Earth)
13.75-14.5	(Earth-to-space)	(Earth-to-space)	(Earth-to-space)
17.2-21.2	(space-to-Earth)	(space-to-Earth)	(space-to-Earth)
27.5-31	(Earth-to-space)	(Earth-to-space)	(Earth-to-space)

A number of projects, which can be found in detail in [3], have been funded for the realization and the planning of emergency communications, many of which contain communications over satellite. In the next subsection, the WISECOM will be analyzed which is a small-scale transported system and is based on satellite backhauling for emergencies.

3.2. WISECOM

The Wireless Infrastructure over Satellite for Emergency Communications (WISECOM) has been co-funded by the European Commission as a solution for the emergency communications based on satellite backhauling [6].

The WISECOM is a telecommunications solution for communications at the "emergency-response" and the "recovery" phase and they have been proposed two versions. The main differences between the two versions are the used satellite systems for connecting the rescue teams with the terrestrial networks not affected by the disaster and the communications systems integrated into each version.

In the first version of WISECOM, the Inmarsat Broadband Global Area Network (BGAN) is employed as the satellite backhauling network of GSM and Wi-Fi networks. The equipment for the first phase is easily transportable by one person and can be very easily deployed within minutes by non-technicians. As it is explained in [6], the advantage of the first version is its capability for easy and rapid development.

The services that can be offered are voice calls between the user terminals even in public telephone networks, short messaging services. Both are essential for the coordination of the rescue teams. Regarding the Wi-Fi networks, they will be available to victims with restricted access to victims and free access to the rescue teams. The system architecture of this first version is shown in Figure 3, as this is given in [6]. The reason why such a version is necessary is that after the disaster, a communication system is needed as quickly as possible in order to satisfy the needs for coordination of rescue teams and the help of the victims.

After the establishment of the basic communication services, there is a need for higher bandwidth and video teleconferences, high data rate transmission of images should be supported. As indicated in [6], robustness and reliability are the biggest requirements for the second phase.

In the WISECOM's second phase the Digital Video Broadcasting- Return Channel via Satellite (DVB-RCS) has been used as a backhauling satellite network. The second version of the system can be deployed in 15-30 minutes and transported by a 4-wheel vehicle or a helicopter.

In this second version, a 2G/3G wider network coverage is offered and technologies like GSM, Wi-Fi/WiMAX for higher bandwidth requirements are supported. The architecture of such a system is shown in Figure 4, as this is taken from [6].

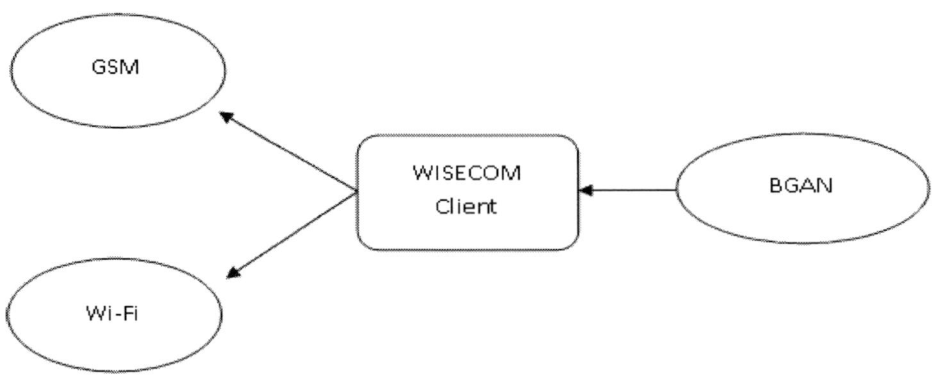

Figure 3. System architecture of the first version of WISECOM.

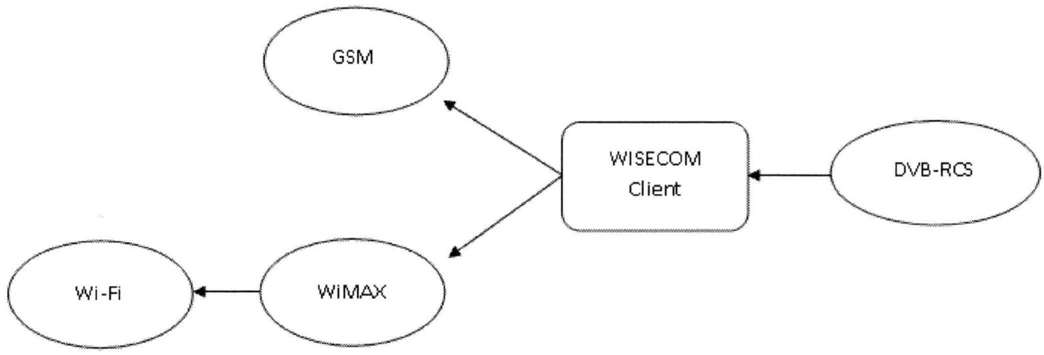

Figure 4. System architecture of the second version of WISECOM.

4. Propagation Impairments and Propagation Modeling

It is obvious from the previous sections that the telecommunications satellites are mostly used for backhauling of the remained terrestrial networks and that generally high bandwidth is required for the emergency communications services. In order to offer high data rates, higher carrier frequencies must be used for the operating system.

As shown in Table 2, ITU has already recommended the allocation of frequencies above 10 GHz and 20 GHz for fixed satellite systems in case of emergency.

In these high frequency bands, the propagated electromagnetic waves are affected by tropospheric phenomena. Precipitation, atmospheric gases, clouds and fog cause the attenuation of the power of the signal, while tropospheric turbulence causes the scintillation of its amplitude [7]. From these atmospheric phenomena and for an operating frequency at Ku and Ka band, rainfall rate is the most severe cause of degradation of the quality of the link.

4.1. Gaseous Absorption

The atmospheric gases that mostly influence the signal in the frequencies of interest, close to 20 GHz, are the oxygen and water vapor. Oxygen and water vapor molecules affect the electromagnetic waves and cause the attenuation of the signal [8]. Liebe's Millimeter-Wave Propagation Model (MPM) [9] is the most accurate model for calculating the gaseous absorption. In this model, the atmosphere is divided into homogeneous layers and for each layer the specific attenuation due to water vapor and oxygen are calculated. The attenuation due to gases is the sum of the specific attenuation for the path on each layer. ITU-R proposes two models for the prediction of attenuation due to atmospheric gases in the range of frequencies less than 350 GHz. Both models refer to oxygen and water vapor absorption and the first one is based on Liebe's methodology [9]. According to this first model, the specific attenuation due to atmospheric gases is shown in Figure 5. Both models require vertical profiles of pressure, temperature and water vapor density, which can be obtained from annual reference models of ITU-R. P. 835 [11].

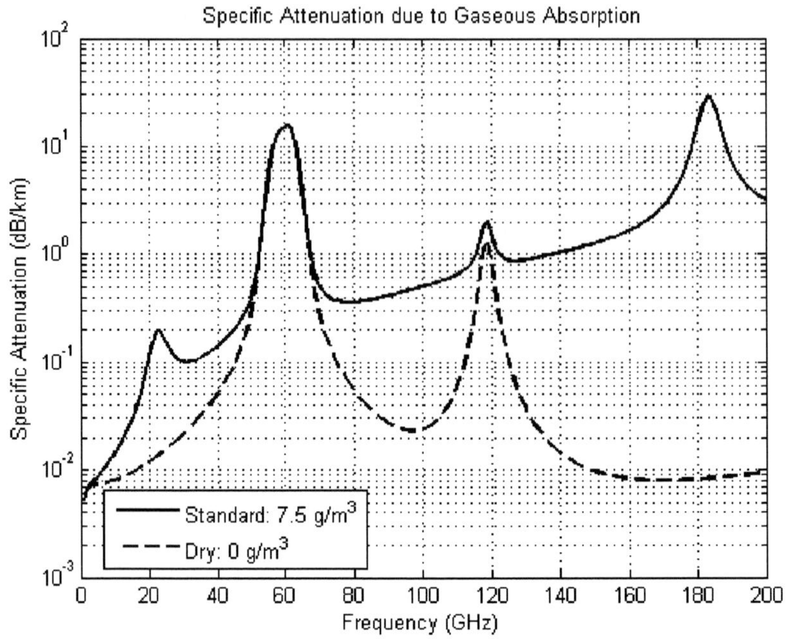

Figure 5. Specific attenuation due to oxygen and water vapor at standard atmosphere and dry air.

However, these models are quite complicated and therefore, ITU-R has recommended in [10] a second model which can be used as an approximation. In this second model approximation algorithms are used as well as surface meteorological data and maps of integrated water vapor content for various time percentages derived by ITU-R. P. 836 [12] for the calculation of the gaseous absorption.

4.2. Clouds Attenuation

Clouds are containing liquid water particles whose size is less than the wavelengths considered in our case.

Rayleigh approximation may be used, as this is recommended in [13], for the calculation of clouds attenuation. According to this model, the specific attenuation due to clouds (γ_c) can be obtained from (1):

$$\gamma_c = K_l M, \quad \text{dB/km} \tag{1}$$

where K_l is the specific attenuation coefficient ((dB/km)/(g/m^3)) and M is the liquid water density in the clouds (g/m^3).

The expressions for calculating the specific attenuation coefficient are given in [13] and it depends on the dielectric permittivity of the water and the temperature. Considering temperature equal to 0°C, the specific attenuation coefficient for clouds can be obtained and it is shown in Figure 6 for various frequencies.

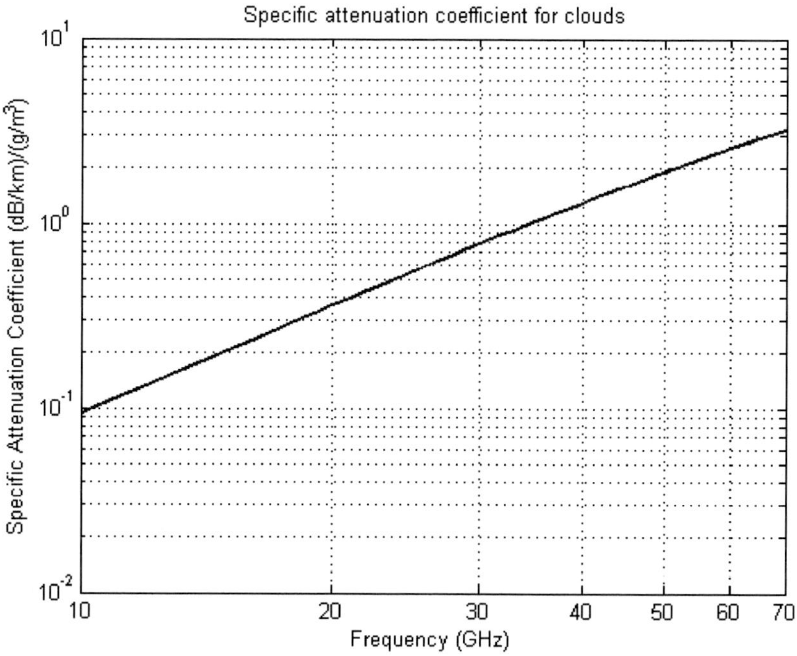

Figure 6. Specific attenuation coefficient vs. frequency.

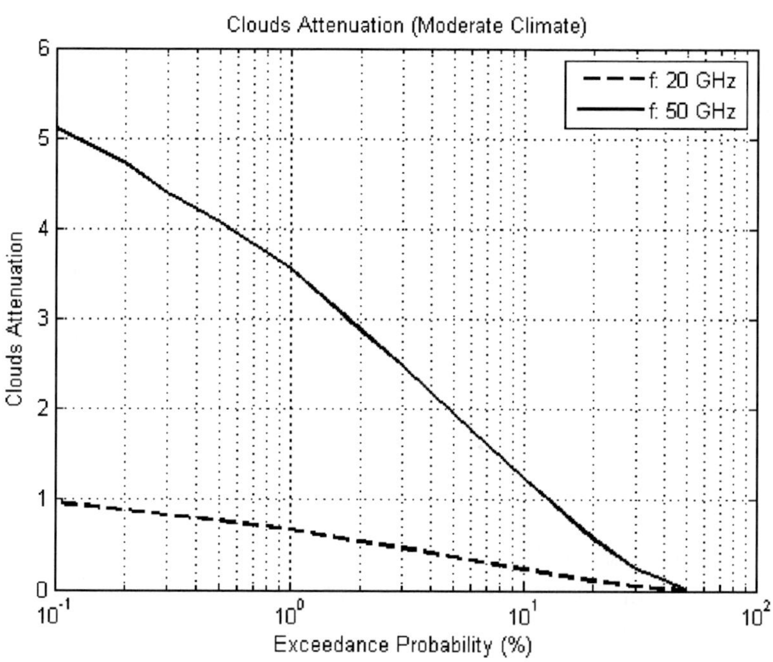

Figure 7. Clouds Attenuation for a moderate climate.

In order to compute clouds attenuation, in [13] the following expression is used:

$$A = \frac{K_l L}{\sin\theta}, \quad 5^o \le \theta \le 30^o \tag{2}$$

where K_l is shown in Figure 6, θ is the elevation angle of the link and L is the total columnar liquid water content and can be derived from the maps of ITU-R. An example of the exceedance probability of clouds attenuation for frequency equal to 20 GHz and 50 GHz for moderate climates is shown in Figure 7.

4.3. Rain Attenuation

Rainfall is the most dominant fading mechanism for the frequencies of interest of this chapter. Due to the scattering and droplet absorption, rainfall causes the attenuation of the signal. The most approved expression for calculating the specific rain attenuation (A_o, dB/km) is the power law, which relates the rain rate (R, mm/h) with rain attenuation [14]:

$$A_o = aR^b \tag{3}$$

where a and b are two coefficients which depend on the elevation angle, frequency and polarization [15]. In this case the rain attenuation along the radiopath can be calculated as:

$$A = \int_0^L A_0(x)dx \tag{4}$$

where L is the length of the path under rain.

Since, rainfall rate has a probabilistic nature rain attenuation can be easily seen as a random variable in order to be able to model it. For the outage prediction of an Earth-space system, the exceedance probability, i.e. the Complementary Cumulative Distribution Function (CCDF) of rain attenuation must be calculated. There are numerous models in the literature for computing the exceedance probability of rain attenuation.

Most of the propagation models require the long-term statistics of rain rate, which can be obtained either from measurements or from the ITU-R Recommendation P. 837 [16]. In Figure 8, the CCDF of rain rate derived from ITU-R. P. 837 is shown for two arbitrary regions one with moderate and one with tropical climate. As it is expected in tropical regions, higher rainfall rate values occur.

We could classify the propagation models into a) empirical, b) statistical and c) semi-physical models. The most known empirical model is the ITU-R model in P. 618 [17]. In this model a step-by step algorithm is given in order to compute the CCDF of rain attenuation given the electrical and geometrical characteristics of the satellite link as well as the rain height. As for the rain rate the value which is exceeded for 0.01% of time is needed. In Figure 9, the CCDF of rain attenuation calculated by the ITU-R model in [17] is shown for an arbitrary region with a moderate climate for two frequencies, 20 and 40 GHz.

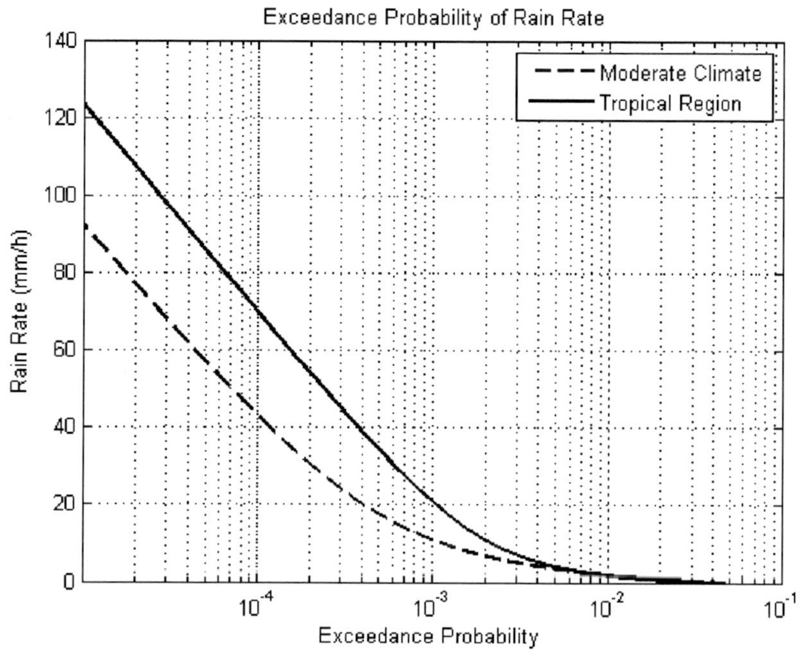

Figure 8. Exceedance probability of rain rate calculated with ITU-R. P. 837.

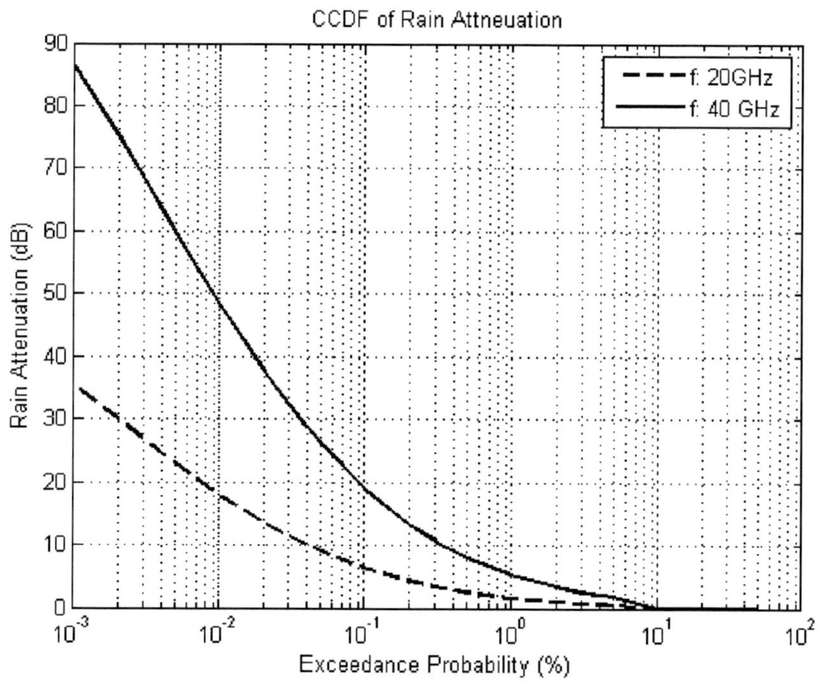

Figure 9. CCDF of rain attenuation.

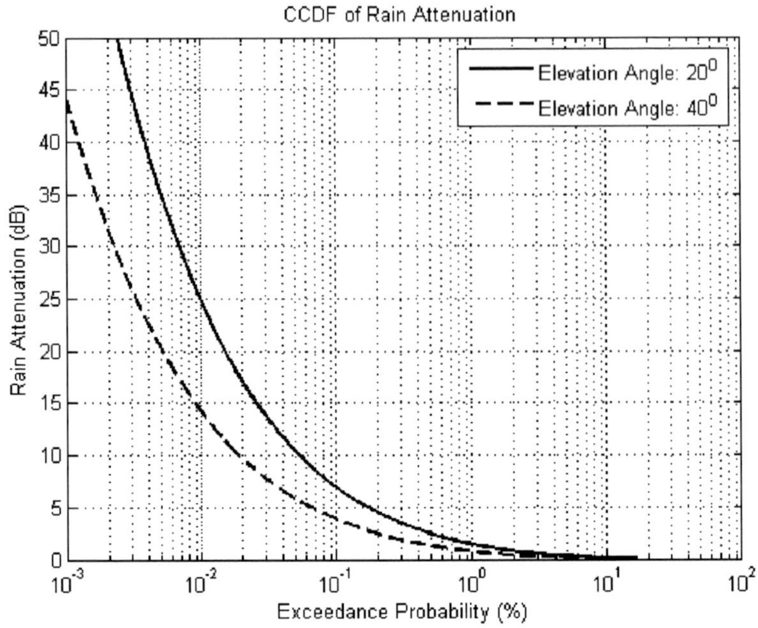

Figure 10. CCDF of rain attenuation.

For the statistical models, rain attenuation and rainfall rate are assumed to follow specific probability distributions. Then adopting a spatial model for rainfall rate or specific rain attenuation, the statistical parameters of rain attenuation are calculated in terms of the statistical parameters of rain rate. A well accepted statistical model is presented in [18]. In this methodology, rainfall rate and rain attenuation follow the lognormal distribution and the spatial distribution of specific rain attenuation shown in (5) is used for the calculation of the parameters of lognormal distribution of rain attenuation.

$$\rho_L = \frac{G}{\sqrt{G^2 + d^2}} \tag{5}$$

where G is a constant which can take values from 0.75 km to 3 km and depends on spatial homogeneity of rain rate which depends on the location and d is the distance between two points, x and x_1 of the radio path, i.e. $d=|x-x_1|$.

Another statistical model based on Lin's methodology has been presented in [19]. The difference between this model and Lin's methodology is that the probability to rain on a specific point is equal to 1. Following this assumption the expressions for calculating the CCDF of rain attenuation are different. The CCDF of rain attenuation for a frequency of 20 GHz and two different elevation angles for a region with a moderate climate are calculated based on the methodology of [19] and shown in Figure 10.

For subtropical and tropical regions a different spatial coefficient has been proposed in [20]:

$$\rho_G = \exp(-a_s \sqrt{d}) \tag{6}$$

where a_s take values from 0.2 to 0.3. A statistical model using the spatial coefficient of (6) has been presented in [21] for subtropical and tropical regions. In heavy rain climatic regions gamma distribution seems to be more appropriate for the modeling of rain rate and rain attenuation for the prediction of the exceedance probability of rain attenuation.

As a last category are the semi-physical models. In this category, the exceedance probability of rain attenuation is computed using the concept of rain cells. Rain cells are closed regions in which it rains, i.e. the rainfall rate is greater than 0. In [22], an exponential shape for the horizontal distribution of rain rate inside a rain cell is proposed, i.e.:

$$R = R_M e^{-\frac{\rho}{\rho_0}} \tag{7}$$

where R_M is the peak rain rate inside the cell, ρ is the distance from the peak and ρ_0 is the distance from the center of the cell in which the rain rate becomes $\frac{R_M}{e}$. The EXCELL model has been used for the prediction of the rain attenuation [23].

However, in [24], the Lowered EXCELL model has been introduced as an improvement to the initial EXCELL model. According to [24], using a probability of occurrence of a rain cell with specific radius and maximum value and knowing the distribution of rain rate, the CCDF of rain rate for stratiform and convective rain can be computed. Based on this methodology and adopting a model for the computation of the effect of melting layer, the CCDF of rain attenuation can be computed. The model is named Stratiform-Convective Excell and has been presented in [25].

Finally, based on the methodology of the EXCELL modeling of rain cells a method for generating two-dimensional rain fields is proposed in [26].

Another semi-physical model is the methodology of HYCELL [27], [28], [29]. According to HYCELL, the horizontal structure of rain rate inside a rain cell can be described by the combination of a Gaussian and an exponential component:

$$R(x,y) = \begin{cases} R_G \exp\left[-\left(\dfrac{x^2}{a_G^2} + \dfrac{y^2}{b_G^2}\right)\right] & , \text{ if } R \ \geq R_1 \\ R_E \exp\left[-\left(\dfrac{x^2}{a_E^2} + \dfrac{y^2}{b_E^2}\right)^{1/2}\right] & , \text{ if } R_2 \leq R < R_1 \end{cases} \tag{8}$$

where R_G is the peak of the rain cell, a_G and b_G is the distances at the x- and y-axes, respectively, in which the rain rate is $\frac{R_G}{e}$ and R_E, a_E and b_E are the equivalent for the exponential component. According to the HYCELL methodology, in order to simulate two-dimensional rain fields, the size of the simulation area, the fractional area over which it is raining and the conditional exceedance probability of rain rate are required. The simulated rain field results from the conglomeration of rain cells inside the observation area. In order to put the rain cells inside the observation area, a random walk is followed. First results of the

HYCELL methodology for calculating the CCDF of rain attenuation have been presented in [30].

4.4. Scintillation

Scintillation of the amplitude of the signal is caused by the tropospheric turbulence, i.e. small variations in time and space of the refractive index of the atmosphere. Scintillation is crucial for low elevation links. One method for calculating the exceedance probability of the scintillation depth is the ITU-R. P. 618 method [17]. According to this method, scintillation depends not only on the radio-electric and geometric characteristics of the link but also on the wet term of radio refractivity. It has been tested to experimental data with a range of frequencies from 7 to 14 GHz. However, its use is recommended for frequencies of at least up to 20 GHz. Moreover, the elevation angle must be greater than 4°. An example is given in figure 11 for a frequency of 20 GHz and two different elevation angles 10° and 40°. The region was selected inside Europe. It can be observed that increasing the elevation angle the scintillation depth is decreasing while the opposite holds for frequency, that is with increasing frequency scintillation depth is increasing.

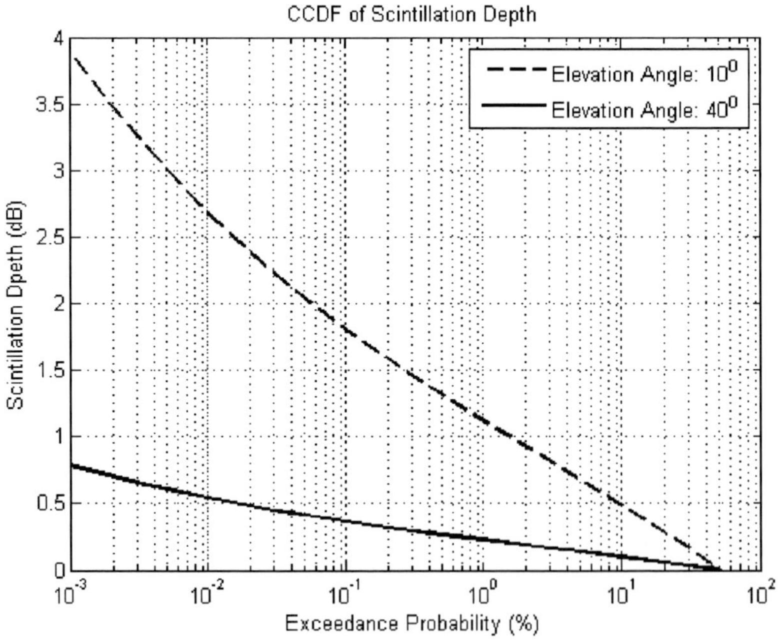

Figure 11. CCDF of scintillation depth.

In [31], another model is proposed for calculating the CCDF of scintillation depth. In this model, radiosonde data are used in order to obtain vertical profiles of temperature, pressure, humidity, wind speed and wind direction. In [31], it is assumed that the probability density function (PDF) of the log-amplitude (x) of the signal can be computed by the following expression:

$$p_l(x) = \int_0^{+\infty} p_s\left(x \mid \sigma_x^2\right) p\left(\sigma_x^2\right) d\sigma_x^2 \tag{9}$$

where σ_x^2 is the variance of the log-amplitude which is assumed to follow the log-normal distribution, given in (10) and $p_s(x \mid \sigma_x^2)$ is the short-term pdf of x which follows the Gaussian distribution, given in (11).

$$p(\sigma_x^2) = \frac{1}{\sqrt{2\pi\sigma_x^2 s}} \exp\left(-\frac{\left[\ln\left(\sigma_x^2 / m\right)\right]^2}{2s^2} \right) \tag{10}$$

with m and s the mean value and standard deviation of the natural logarithm of σ_x^2.

$$p_s(x \mid \sigma_x^2) = \frac{1}{\sqrt{2\pi\sigma_x^2}} \exp\left(-\frac{x^2}{2\sigma_x^2} \right) \tag{11}$$

Then relating the variance of the scintillation with the structure parameter (see (12)) and using the mean structure parameter and its relationship with the parameters of temperature, humidity, pressure and wind as this is given in [32], the CCDF of scintillation depth can be calculated. However, for this method radiosonde data are needed.

$$\sigma_x^2 = 42.48 \frac{k^{7/6}}{(\sin\theta)^{11/6}} \int_{height} C_n^2(z) z^{5/6} dz \tag{12}$$

where θ is the elevation angle of the link, k is the wavenumber and $C_n^2(z)$ is the structure parameter of the refractive index n at height z.

4.5. Total Attenuation

Although various models exist for the computation of the exceedance probability of each attenuation component, the combination of these phenomena for calculating the exceedance probability of total attenuation is a difficult subject. Especially when the availability which must be guaranteed is low, apart from rain attenuation all the other effects must be taken into account. There are various methods for combining these effects considering if the effects are fully correlated or uncorrelated [33]:

1. Equiprobability Summing

In this method the attenuation predicted at a certain probability level *P*% are added. Therefore, if $A_1(P)$ and $A_2(P)$ are two attenuation values at *P* percentage of time then the total is:

$$A_{tot}(P) = A_1(P) + A_2(P) \tag{13}$$

In this case the two phenomena are considered full correlated and so it is the worst case estimation.

2. Convolution Method

In this case the two phenomena are considered uncorrelated the convolution method may be used. So the cumulative distribution of the total attenuation is calculated from:

$$P(A_t) = \int_{-\infty}^{A_t} p_1(A) P_2(A_t - A) dA \tag{14}$$

where $p_1(A)$ is the PDF of attenuation due to phenomenon 1 and $P_2(A)$ the cumulative distribution of attenuation due to phenomenon 2.

3. Disjoint Summing

The exceedannce probabilities of two phenomena can be added if these are disjoint. So, for example in case that the CCDF of attenuation in rainy and non-rainy time are available, the CCDF of total attenuation could be:

$$P(A_t > A) = P(A_{non-rain} > A) + P(A_{rain} > A) \tag{15}$$

However, to estimate the CCDF of disjoint attenuation due to different propagation phenomena is difficult.

4. Root-Sum-Square Summing

In this case the total attenuation is computed according to the following expression:

$$A_{tot}(P) = \sqrt{A_1^2(P) + A_2^2(P)} \tag{16}$$

Equation (16) underestimates the total attenuation value in comparison to the equiprobability summing.

5. Coherent Summing

If the attenuation values are known simultaneously at a given time t_j then the total attenuation is computed by:

$$A_{tot}(t_i) = \sum_{j=1}^{n} A_j(t_i) \tag{17}$$

Therefore, if the attenuation components are known at a given period of time the distribution of the total attenuation can be calculated.

For the computation of total attenuation there are various methods which can be found in [33]. Among them is the ITU-R method which has been recommended in ITU-R. P.618 [17]. According to this methodology, the total attenuation can be calculated using the following expression:

$$A_T(P) = A_G(P) + \sqrt{\left(A_R(P) + A_C(P)\right)^2 + A_S^2(P)} \tag{18}$$

where A_T, A_G, A_R, A_C and A_S are the total attenuation, the gaseous attenuation, the rain and cloud attenuation and the scintillation fade depth, respectively, at a given probability. In figure 12, the total attenuation has been computed following ITU-R. P. 618 for a region with a moderate climate and a frequency of 20 GHz for two different elevation angles 20° and 40°.

4.6. Time Series Generators

As it will be shown in the next section of this chapter, Fade Mitigation Techniques (FMTs) are needed in order to offer high availability satellite services and high data rates. FMTs are control loops that compensate in real time for propagation impairments due to fading (rain attenuation in principle). Rain attenuation Time series Generators are used in order to simulate FMT control loops and assess their performance.

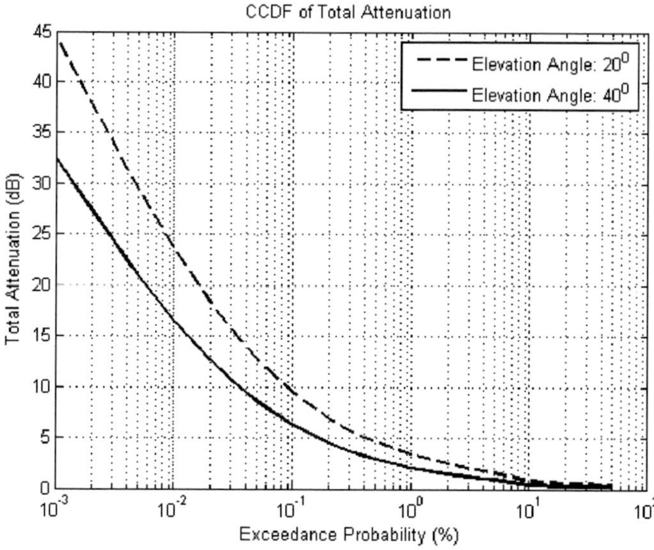

Figure 12. CCDF of total attenuation.

Such Time Series Generators to be reliable, should reproduce the statistical properties of the phenomena they are modeling In most of the cases, reproducing first order (PDF, CCDF) and second order (Spectral Density Function, Autocorrelation) gives a very reliable Generator.

Since, rain attenuation gives the highest attenuation values and has the greatest space-time variability, most time series generators refer to rain attenuation.

More particularly, Maseng and Bakken in [34] have presented a time series generator based on stochastic differential equations (SDEs), assuming that rain attenuation follow the lognormal distribution and that it is a continuous-time Markov process. The SDE which describes the rain attenuation (A_t) is:

$$dA_t = A_t\beta[\sigma_\alpha^2 - \ln(A_t / A_m)]dt + \sqrt{2\beta}A_t\sigma_\alpha dW_t \quad (19)$$

where A_m and σ_α^2 are the median value of rain attenuation and the variance of the natural logarithm of rain attenuation, β is the dynamic parameter of the model and W_t is the Brownian Motion [35]. According to (19), rain attenuation follows the lognormal distribution, the rate of change of rain attenuation is proportional to the initial value of rain attenuation and the autocorrelation function can be calculated with the following expression:

$$R_A(\tau) = A_m^2 \exp(\sigma_a^2(1 + \rho(\tau))) \quad (20)$$

where

$$\rho(\tau) = \exp(-\beta\tau) \quad (21)$$

which is the autocorrelation function of the reduced variable $X_t = \frac{\ln(A_t / A_m)}{\sigma_a}$, which follows the Gaussian distribution. The reduced process has the following power spectrum:

$$W(f) = \frac{1}{1 + \left(\frac{2\pi f}{\beta}\right)^2} \quad (22)$$

The simulation of the time series of rain attenuation based on (19) can be realized following the diagram presented in Figure 13. First, a white noise with mean 0 and standard deviation 1 ($n(t)$) is generated and passes through a low-pass filter with a cut-off frequency β.

The output of the filter is the Gaussian process X_t. Then transforming this process with the transformation $A_t = A_m \exp(\sigma_a X_t)$, time series of rain attenuation can be generated. An example of rain attenuation time series simulated with the Maseng-Bakken model during a rain event is shown in Figure 14 for two different frequencies.

The generation of rain attenuation series in [34] is based on a numerical solution implemented with filters. In [36], the continuous-time analytical solution has been obtained and presented.

From (19) using the Ito's formula [35], the SDE which describes the random variable X_t can be obtained. As it has been shown in [34], this SDE is the Langevin equation [35]:

$$dX_t = -\beta X_t dt + \sqrt{2\beta} dW_t \tag{23}$$

The continuous-time solution of (23) is shown in [35]:

$$X_t = \exp(-\beta t)\left(X_0 + \sqrt{2\beta} \int_0^t \exp(\beta s) dW_s \right) \tag{24}$$

where X_0 is the initial value of random process X_t. Then the rain attenuation time series can be derived from X_t using the transformation $A_t = A_m \exp(\sigma_a X_t)$.

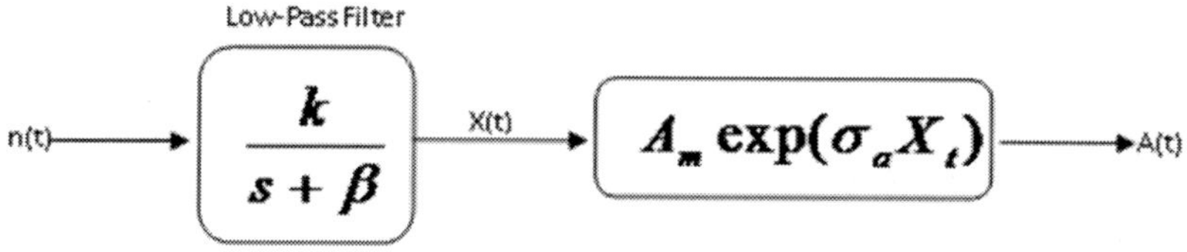

Figure 13. Block diagram for generating rain attenuation.

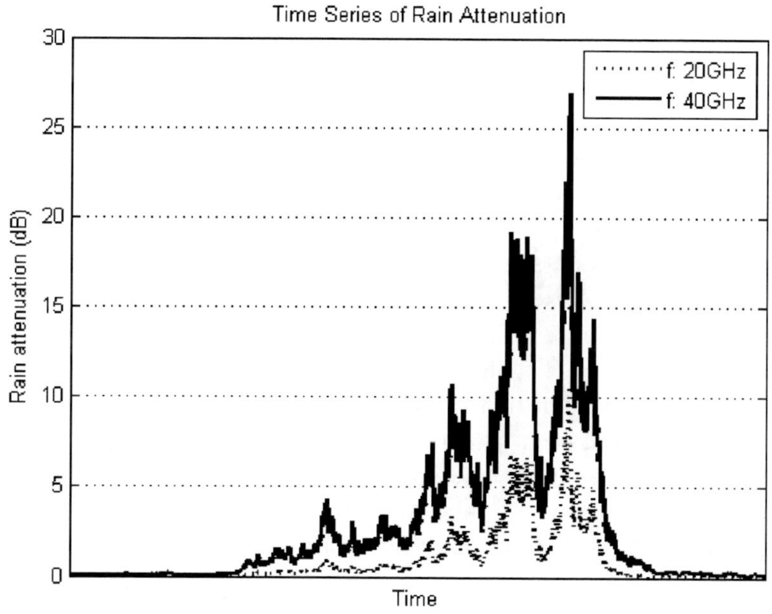

Figure 14. Rain attenuation time series.

The derivation of the input parameters of the Maseng-Bakken model is very important for the time series generations. First, the statistical parameters (A_m and σ_α) of the lognormal distribution of rain attenuation can be obtained through a regression process, fitting the CCDF of rain attenuation, either experimental or a predicted one, with the CCDF of lognormal distribution. For the derivation of the dynamic parameter β, more than one methods have been identified. In [36], the theoretical and experimental mean value of the hitting time distribution have been used in order to obtain the dynamic parameter. The definition of the diffusion term of the SDE, given in (19) is:

$$\frac{E\left[\left(A(t+\Delta t)-A(t)\right)^2 \mid A(t)\right]}{\Delta t}=2\beta A^2\sigma_a^2 \tag{25}$$

where E[ ] is the expectance operator. This expression has been taken in [37] in order to derive the dynamic parameter from experimental data, while in [38] the dynamic parameter is calculated by fitting the spectrum of X_t (22) and the autocorrelation function (21) to the experimental ones.

The time series synthesizer for rain attenuation which is recommended by ITU in ITU-R. P. 1853 [39] is based on the Maseng-Bakken model with the difference that the statistical parameters of rain attenuation refer to the conditional rain attenuation ($P(A \geq A_{th} \mid A > 0)$) and an offset parameter was introduced in order to take into account the probability that rain attenuation exceeds the 0 dB value ($P(A>0)$). Based on the Maseng-Bakken model, a rain event synthesizer has been developed, which is presented in [40]. In [41], Matricciani uses the two-layer model in the Synthetic Storm Technique (SST) in order to derive time series of rain attenuation given the time series of rain rate. This method can be also used for the derivation of the first-order statistics of rain attenuation. However, it requires the time series of rain rate which can be derived from measurements and the average speed of the wind. In Figure 15, rain attenuation time series given as input rain rate time series measured at Athens [42] is shown. More time series synthesizers can be found on [43] and [44] which they also include extensive comparison.

As it was stated in the total attenuation modeling, in low availability systems, gaseous and clouds attenuation as well as scintillation fade depth are also crucial for system design. For the above reason, time series prediction models are very important for each attenuation component. Very recently, in ITU-R P. 1853-1 [39] there have proposed time series synthesizers for all the attenuation components and their combination in order to obtain total attenuation time series. The rain attenuation synthesizer in [39] has been previously described. For clouds attenuation time series, it is recommended a model based on filtering method and similar to the one used for rain attenuation.

According to [39], the integrated liquid water content (ILWC) is following lognormal distribution. Generating Gaussian noise and then filtered by two parallel low-pass filters similar to these of the Maseng-Bakken model with two different cut-off frequencies, correlated Gaussian noise can be derived.

As a next step the time series are calibrated in order to take into account the probability of cloud occurrence and then they are transformed into lognormal random variables. In Figure 16 cloud time series are shown as these are generated from [39].

As for the gaseous attenuation, oxygen attenuation is assumed constant and the water vapor absorption is computed with the approximation algorithm in [10]. In order to obtain time series of attenuation due to water vapor first time series of integrated water vapor content are generated and then these are used for the computation of water vapor absorption.

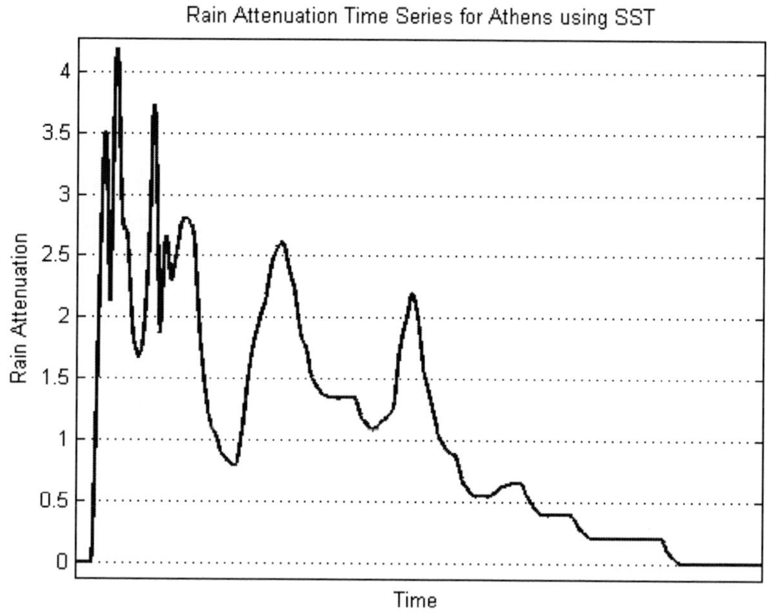

Figure 15. Rain attenuation time series generating with SST, from rainfall rate data in Athens.

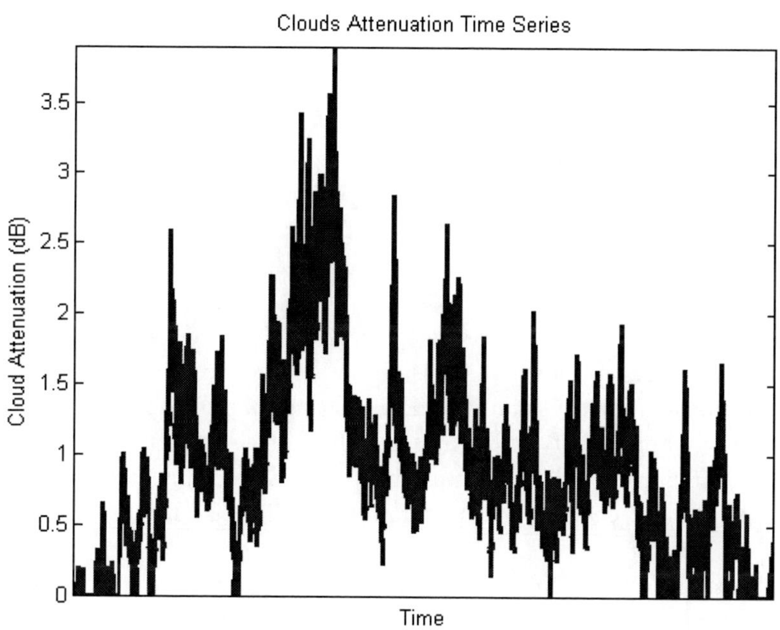

Figure 16. Clouds attenuation time series from ITU-R P.1853-1.

The methodology of the time series synthesizer for IWVC presented in [39] is based on the filtering of Gaussian white noise from a low-pas filter and then the transformation of the Gaussian process into a process which follows the Weibull distribution. Considering scintillation time series, white Gaussian noise is filtered through a low-pass filter with a cut-off frequency 0.1 Hz. The filter must be designed in a way that the asymptotic power spectrum of the time series has a $f^{8/3}$ roll-off.

Finally, in order to generate total attenuation time series, the white Gaussian noise which is used as an input to the rain attenuation synthesizer and the ILWC synthesizer is the same, while the one used for the IWVC and ILWC have a correlation coefficient 0.8. The scintillation depth time series apart from the scintillation time series depend on the IWVC time series and the rain attenuation time series. With increasing rain attenuation, scintillation depth increases. The total attenuation time series are derived from the coherent method, i.e. from the addition of each component at the same time instance. As a last point for the propagation modeling is the space-time field generators of meteorological metrics which affect the attenuation components, in order to have spatially and temporally correlated atmospheric attenuation. Although there are models for simulating spatial rain fields, such as HYCELL and multiEXCELL models and spatially correlated cloud fields [45], these do not take into account the temporal correlation. Models which are capturing the spatial and temporal correlation for rain attenuation have been presented in [46], [47], [48], while total attenuation space-time fields can be generated using the methodology presented in [49].

5. Fade Mitigation Techniques

As it was observed in the previous section total attenuation and mainly rain can exceed very high attenuation values for a small time percentage. Therefore, for a system which requires high availability the attenuation values exceeded for small time percentages are crucial. One solution in order to deal with the attenuation is to employ a fade margin. However, due to the high values of the fade margin which are required to compensate attenuation which is exceeded for a small time percentage, the sources of the system will remain unexploited for the most of the period of operation. So, in order that a system be efficient and operate under low fade margin, Fade Mitigation Techniques (FMTs) are required.

FMTs can be classified into three categories: a) Effective Isotropic Radiated Power (EIRP) control techniques, b) adaptive transmission techniques and c) diversity techniques. Most of the FMTs use a control loop in order to change their state. According to this loop, first the observation of the quality of the link must take place and then the state of the satellite channel and its duration on this state must be predicted. Finally, after the prediction and estimation of the satellite channel and its duration the FMT is enabled in order to adjust its parameters.

The identification of the channel can be achieved by the measurements of the channel or by a short-term prediction. For the short-term prediction, the statistics of fade slope may be used. There are a number of models for the modeling of the long-term statistics of the fade slope, a review on which can be found in [8] and [44]. The Van de Kamp model, presented in [50] has been adopted by ITU in ITU-R. P. 1623 [51]. Another method for the short-term prediction of fades at Ka band has been presented in [52] and it is assumed that attenuation is an autoregressive moving average process. As for the duration of the fades, the fade duration statistics are used, for which various models have been developed and can be found in [8] and [44]. Considering fade duration statistics, these can be divided into short fades and long fades. The model of ITU-R. P. 1623 [51] models long fades with a lognormal distribution and the short fades with a power law. Fade slope and fade duration can be also found in literature as fade dynamics. Fade dynamics depend on first and second-order statistics and can be also modeled with the time series generators, presented in the previous section.

Another metric which may be proven essential for the FMTs is the hitting time distribution. Hitting time is the time at which a stochastic process reaches one value of the boundary (A_{min}, A_{max}) given that the initial value is A_0 with $A_{\min} < A_0 < A_{\max}$. Hitting time distribution reflects the dynamics of the process, given the SDE that describes it. For rain attenuation and the Maseng-Bakken model, the hitting time distribution has been presented in [36]. In figure 17, the CCDF of hitting time distribution for a link in Europe with an operating frequency of 20 GHz is given. In the case of the figure the A_{max} was set infinite and so the hitting time distribution is the time needed for the process to reach A_{th}=A_{min} starting from A_0. Hitting time distribution can be used for the estimation of the time needed for the FMT to change its state [53].

In the EIRP control techniques, in case of fades the power of the Earth station or the satellite can be changed or the antenna gain. In case of the adjustment of power at the Earth station, the technique is called Uplink Power Control (ULPC) while at the satellite Downlink Power Control (DLPC). The two drawbacks in the technique of ULPC is the adjacent channel interference due to the increased power sent at the sidelobes and the adjacent satellite interference.

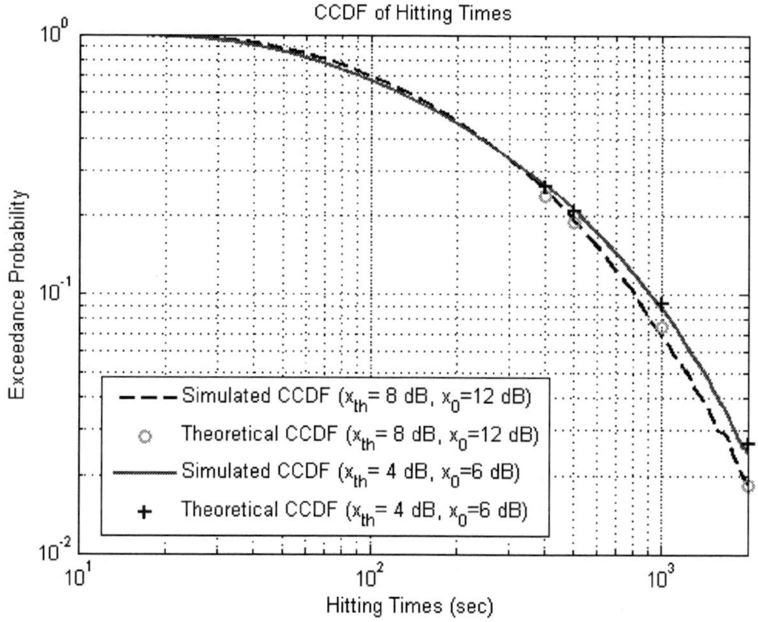

Figure 17. CCDF of hitting times.

The adjacent satellite interference occurs due to the increased power that it is received by the satellites which are close to the receiver satellite. On the other hand, the implementation of DLPC is more difficult due to the satellite size and weight. Also, in DLPC adjacent channel interference may occur. Two more drawbacks are the intermodulation interference and the intersystem interference.

Intermodulation interference is caused by the nonlinear amplification of the multiple carriers, while intersystem interference is caused by the interference of the Earth-space system to terrestrial networks due to the increased power.

Another EIRP control technique is the spot beam shaping by the adjustment of the antenna gain. A Geostationary satellite which covers a wide region may provide services using spot beam in order that the antenna beamwidth is reduced and the gain is increased. In this technique short-term weather predictions are needed which may be derived from satellite images or numerical weather products. Using these prediction the antenna gain may be changed in the region which is affected more by the atmospheric phenomena at the specific time instance. Given the meteorological conditions the optimization of the spot beams may be static or dynamic [54].

The adaptive transmission techniques include the adaptive coding (AC), adaptive modulation (AM) and the data rate reduction (DRR). In AC technique bits are added into the transmitted symbol in order to detect and correct erroneous symbols.

However, due to the increased size of the symbol the bandwidth have to be increased in order to keep the data rate the same. There are various coding schemes, as this is analyzed in [7]. In case of short fades such as scintillation, the interleaving method is proposed. In case of strong fades complex coding schemes must be used. In AM technique the current modulation which is used is changing in order to have reduced BER for a given probability error. In this case, the spectral efficiency is reduced and so the data rate is reduced. In clear sky conditions highly efficient modulation schemes may be used, but under rain fading conditions schemes such as BPSK and QPSK shall be used. The last technique in the adaptive transmission techniques is DRR. In DRR, the data rate is decreased while the BER threshold and the bandwidth remain constant and so a reduced carrier –to-noise threshold is required.

The last category of the FMTs is the diversity techniques. The diversity techniques are: the site diversity, orbital (satellite) diversity, frequency diversity and time diversity.

In Figure 18 a dual-site diversity system is shown. In these techniques, two Earth stations are communicating with the same satellite and the Earth-space path with the least attenuation is selected. This technique takes advantage of the fact that as the distance between two points is increased, the decorrelation of rain attenuation is also increased. Since, site diversity is the most efficient FMT model for the counteraction of rain attenuation, models which estimate the joint exceedance probability, $P(A_1 > A_{thr},\ A_2 > A_{thr})$, and site diversity gain have been developed [17], [55], [56], [57], as well as time series of rain attenuation correlated in space [46], [47]. In Figure 19, time series of rain attenuation correlated in space are shown for separate Earth-space stations, as well as the attenuation for the dual site diversity system. In orbital diversity instead of two Earth stations, there is only one Earth station (probably implemented with two receiving antennas) which communicates with two satellites, as shown in Figure 20, and as in site diversity the path with the least attenuation is selected. When the angular separation between the two radiopaths is increasing the decorrelation of rain attenuation also increases. In comparison to site diversity, it is less cost inefficient, since it does not require the installation of one more Earth station. However, it is less effective. Attenuation due to the atmospheric phenomena increases as the frequency increases.

This holds especially for rain attenuation in the operating frequencies of interest in Emergency communications. Therefore, in case that high attenuation values are observed, the system may operate at low frequencies while in clear sky conditions at high frequency. The last technique is time diversity technique. It takes advantage of the increased decorrelation of rain rate with time. Therefore, when in a time instance the attenuation values are high, the system may retransmit the same symbols when the meteorological conditions are such that the fades are weak.

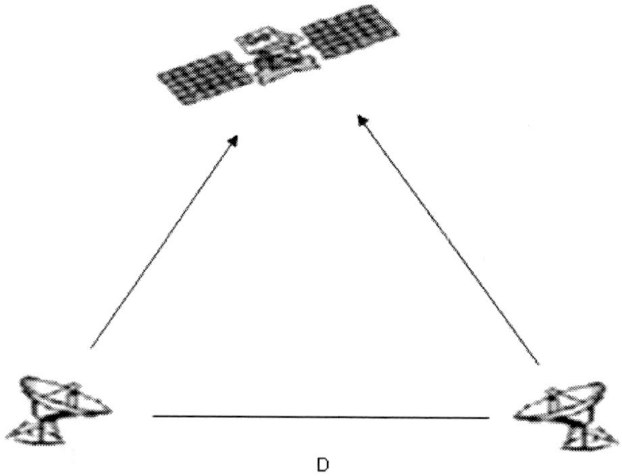

Figure 18. Dual-site diversity system.

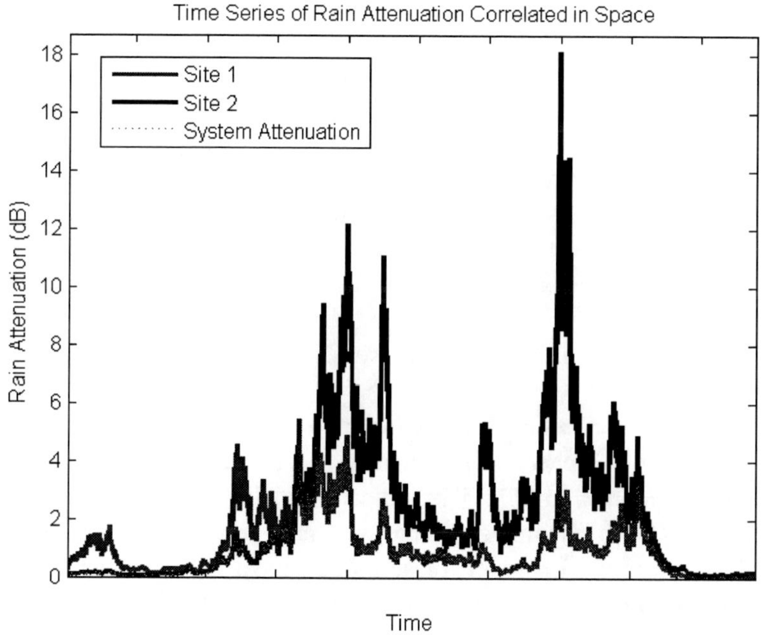

Figure 19. Rain Attenuation time series correlated in space for two sites and the attenuation for the dual site diversity system (selection combining).

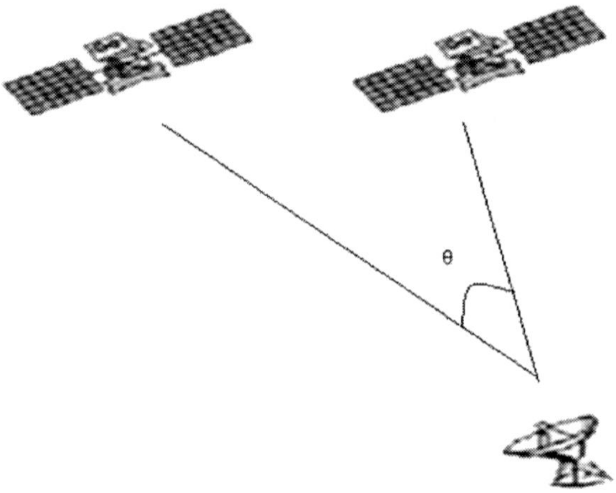

Figure 20. Orbital diversity system with two satellites.

The time needed for the retransmission can be modeled with the concept of hitting time statistics, while the dynamics of attenuation are modeled with time series. In [58] and [59], two models are presented for the computation of joint exceedance probability of rain attenuation for two different time instances.

6. Future Research Directions

Propagation modeling as it has been referred is crucial for the reliable design of an emergency communication system. On the other hand, for the evaluation of the FMTs and their implementation with control loops require the temporal and spatial properties of the attenuation components and the meteorological parameters which affect them. For emergency communications, the systems must be available at a very high time percentage and they have to offer high data rates. Moreover, in satellite emergency communications, not all the FMTs have the same impact. For example, time diversity in a disaster will not be the most appropriate since time delay is essential while the teams must be coordinated. However, orbital and site diversity techniques could be very effective and efficient given the redundancy of the Earth stations at the far-field. Therefore, a first task for future research is the assessment of the FMTs in emergency communications not only on the system perspective (performance) but also on the services which must be offered during the emergency response.

A second issue for future research is the improvement of the models which predict outage prediction due to the atmospheric phenomena. In telecommunication satellite services at operating frequencies close to 20 GHz, rain attenuation is the most severe cause of degradation of the quality of the link and therefore, many models have been developed for the outage prediction under rain. However, these models have not been evaluated under a large database, to find the most accurate one. Furthermore, in high availability systems such as satellite emergency communications, attenuation due to other than rain atmospheric phenomena must be predicted. Therefore, the evaluation of the accuracy of the present models as well as the development of new models is required. Also, research on the combination of attenuation components in order to obtain the total attenuation exceedance probability is required. The same holds for the dynamics of the attenuation components and the dynamics of total attenuation.

The development of space-time synthesizers is also an open issue on propagation modeling. Spatially and temporally correlated total attenuation fields can be used for the evaluation of large-scale systems in a region of a country or even continent, in case of natural disasters. Moreover, fixed global telecommunications systems which will be able to operate in a continental region could be a solution for emergency communications. Such systems will definitely need the space-time generators of total attenuation fields, since many gateways across the continent or the country will be cooperating and communicating in order to support and coordinate the rescue teams.

Conclusion

In this chapter the propagation issues and propagation modeling for the emergency satellite-based communications are analyzed. First, the communication flow during the emergency response is described, making clear that the communication with high bandwidth is required for the support of the various services during disasters. Since the terrestrial infrastructure may be damaged, satellites will be used for the communication needed at the emergency response phase, for the backhauling of the system. In order to support higher data rates, the emergency satellite systems must operate at higher frequencies at Ku band and above. However, in these high frequency bands, tropospheric phenomena affect the quality of the link. The greatest impact on high availability systems comes from atmospheric gases, precipitation, clouds and tropospheric turbulence. In this Book chapter, these phenomena are described in detail and well-accepted models for the prediction of the first order statistics have been given. In case of the evaluation of the satellite emergency system, the fade dynamics are also important and therefore accurate time series generators for propagation modeling are presented. Furthermore, the FMTs which must be implemented into the emergency satellite system in order to compensate for total attenuation are described and plans for future research are discussed.

Acknowledgment

This work has been done under the framework of THALES-NTUA MIMOSA-Network of Excellence project funded by EU and GSRT.

References

[1] ETSI-EMTEL Working Group- www.emtel.etsi.org

[2] Satellite Earth Stations and Systems (SES); Overview of Present Satellite Emergency Communications Resources, *ETSI Technical Report* 102 641 v1.1.1, 2008.

[3] L. Franck, M. Berioli, P. Boutry, G. Harles, L. S. Ronga, R. Suffritti, L. Thomasson, "On the Role of Satellite Communications for Emergency Situation with a Focus on Europe", International Journal of Satellite Communications and Networking, vol. 29, pp. 387-399, 2011.

[4] ITU-R Emergency Radiocommunication Services, http://www.itu.int/ITU-R/index.asp?category=informationandrlink= emergencyandlang=en

[5] ITU-R RS. 1859, "Use of Remote Sensing Systems for Data Collection to be used in the Event of Natural Disasters and Similar Emergencies", Geneva, 2010.

[6] M. Berioli, J. M. Chaves, N. Courville, P. Boutry, J.-L. Fondere, H. Skinnemoen, H. Tork, M. Werner, M. Weinlich, "WISECOM: A Rapidly Deployable Satellite Backhauling System for Emergency Communications", *International Journal of Satellite Communications and Netwrokin,* 2010.

[7] A. D. Panagopoulos, P-.D. M. Arapoglou, P. G. Cottis, "Satellite Communications at Ku, Ka, and V Bands: Propagation Impairments and Mitigation Techniques", *IEEE Communications Surveys*, Third Quarter, 2004.

[8] ESTEC Contr. 17760/03/NL/JA, "*Characterization and Modeling of Propagation Effects in the 20-50 GHz Band*", Final Report.

[9] H. Liebe, "MPM-An Atmospheric Millimiter-Wave Propagation Model", International Journal of Infrared and Millimeter Waves, vol. 10, no. 6, 1989.

[10] ITU-R. P. 676-9, "Attenuation by Atmospheric Gases", *International Telecommunication Union*, Geneva, 2012.

[11] ITU-R. P. 835-5, "Reference Standard Atmospheres", International Telecommunication Union, Geneva, 2012.

[12] ITU-R. P. 836-4, "Water Vapour: Surface Density and Total Columnar Content", *International Telecommunication Union*, 2009.

[13] ITU-R. P. 840-5, "Attenuation due to Clouds and Fog", *International Telecommunication Union*, Geneva, 2012.

[14] R. L. Olsen, D. V. Rogers, D. B. Hodge, "The aR^b Relation in the Calculation of Rain Attenuation", *IEEE Transactions on Antennas and Propagation,* vol. Ap-26, no. 2, pp. 318-329, 1978.

[15] ITU-R. P. 838, "Specific Attenuation Model for Rain for Use in Prediction Methods", *International Telecommunication Union*, 2005.

[16] ITU-R. P. 837-6, "Characteristics of Precipitation for Propagation Modelling", *International Telecommunication Union,* Geneva, 2012.

[17] ITU-R. P. 618-10, "Propagation Data and Prediction Methods Required for the Design of Earth-space Telecommunication Systems", *International Telecommunication Union,* 2009.

[18] S. H. Lin, "A Method for Calculating Rain Attenuation Distributions on Mcrowave Paths", *Bell system technical journal*, vol. 54, no. 6, pp. 1051-1083, 1975.

[19] J. D. Kanellopoulos, S. G. Koukoulas, "Analysis of the Rain Outage Performance of Route Diversity System", *Radio Science,* vol. 22, no. 4, pp. 549-565, 1987.

[20] K. Morita, I. Higutti, "Prediction Methods for Rain Attenuation Distribution of Micro and Millimeter Waves", rev. *Electr. Commun.* Labs, vol. 54 (7-8), pp. 651-558, 1976.

[21] D. Panagopoulos and J. D. Kanellopoulos "Adjacent Satellite Interference Effects as applied to the Outage Performance of an Earth-Space System located in a Heavy Rain Climatic Region", *Annals of Telecommunications,* No 9-10, pp.925-942, 2002.

[22] Capsoni, F. Fedi, C. Magistroni, A. Paraboni, A. Pawlina, "Data and Theory for a new Model of the Horizontal Structure of Rain Cells for Propagation Applications", *Radio Science*, vol. 22, no. 3, pp. 395-404, 1987.

[23] Capsoni, F. Fedi, A. Paraboni, "A Comprehensive Meteorologically Oriented Methodology for the Prediction of Wave Propagation Parameters in Telecommunication Applications beyond 10 GHz", *Radio Science,* vol. 22, no. 3, pp. 387-393, 1987.

[24] Capsoni, L. Luini, C. Riva, "Stratiform and Convective Rain Discrimination Starting from the Site P(R)", *IEEE Transaction on Antennas and Propagation*, vol. 54, no. 11, pp. 3566-3569, 2006.

[25] C. Capsoni, L. Luini, A. Paraboni, C. Riva, A. Martellucci, "A New Prediction Model of Rain Attenuation that Separately Accounts for Stratiform and Convective Rain", *IEEE Transactions on Antennas and Propagation*, vol. 57, no.1, 2009.

[26] L. Luini, C. Capsoni, "MultiEXCELL: A New Rain Field Model for Propagation Applications", *IEEE Transactions on Antennas and Propagation*, vol. 59, no. 11, 2011.

[27] L. Feral, H. Sauvageot, L. Castanet, J. Lemorton, "HYCELL-A New Hybrid Model of the Rain Horizontal Distribution for Propagation Studies: 1. Modeling of the Rain Cell", *Radio Science*, 2003.

[28] L. Feral, H. Sauvageot, L. Castanet, J. Lemorton, "HYCELL-A New Hybrid Model of the Rain Horizontal Distribution for Propagation Studies: 2. Statistical Modeling of the Rain Rate Field", *Radio Science,* 2003.

[29] L. Feral, H. Sauvageot, L. Castanet, J. Lemorton, F. Cornet, K. Leconte, "Large-Scale Modeling of Rain Fields from a Rain Cell Deterministic Model", *Radio Science*, 2006.

[30] C. Kourogiorgas, L. Feral, V. Fabbro, N. Jeannin, J. Lemorton, "Use of HYCELL Model for Predicting Rain Attenuation on UAS Low Elevation Datalinks", 4th EuCAP, Barcelona, Spain, 2010.

[31] H. Vasseur, "Prediction of Tropospheric Scintillation on Satellite Links from Radiosonde Data", *IEEE Transactions on Antennas and Propagation,* vol. 47, no. 2, 1999.

[32] J. M. Warnock, T. E. VanZandt, "A Statistical Model to Estimate Refractivity Turbulence Structure Constant C_n^2 in the Free Atmosphere", NOOA Tech. MEMO ERL, AL-10, NOOA, Aeronom. Lab., boulder Co, 1985.

[33] COST Action 255, "Radiowave Propagation Modeling for SatCom Services at Ku-Band and Above", *Final Report*, 2002

[34] T. Maseng, P. M. bakken, "A Stochastic Dynamic Model of Rain Attenuation", *IEEE Transactions on Communications*, vol. 29, no. 5, pp. 660-669, 1981.

[35] S. Karlin, H. Taylor, "*A Second Course in Stochastic Processes*", New York: Academic, 1981.

[36] S. A. Kanellopoulos, A. D. Panagopoulos, J. D. Kanellopoulos, "Calculation of the Dynamic Input Parameter for a Stochastic Model Simulating Rain Attenuation: A Novel Mathematical Approach", *IEEE Transactions on Antennas and Propagation*, vol. 55, no. 11, 2007.

[37] G. Carrie, F. Lacoste, L. Castanet, "A Robust Estimator of the Dynamic Characteristics of Rain Attenuation Events", *Proc. of the 2nd JA2310 SatNEx Workshop on Channel Modeling and Propagation Impairments Simulation,* Oberpfaffenhofen, Germany, 2008.

[38] Burgueno, E. Vilar, M. Puigcerver, "Spectral Analysis of 49 Years of Rainfall Rate and Relation to Fade Dynamics", *IEEE Transactions on Communications,* vol. 38, no. 9, 1990.

[39] ITU-R. P. 1853-1, "Tropospheric Attenuation Time Series Synthseis", *International Telecommunication Union*, Geneva, 2012.

[40] G. Carrie, F. Lacoste, L. Castanet, "A New 'Event-on-Demand' Synthesizer of Rain Attenuation Time Series at Ku-,Ka- and Q/V-bands", *International Journal of Satellite Communications and Networking*, pp. 47-60, 2011.

[41] E. Matricciani, "Physical-Mathematical Model of the Dynamics of Rain Attenuation Based on Rain Rate Time Series and a Two-Layer Vertical Structure of Precipitation", *Radio Science*, vol. 31, no. 2, pp. 281-295, 1996.

[42] C. I. Kourogiorgas, A. D Panagopoulos, N. Moraitis, S. N. Livieratos, G. E. Chatzarakis, J. D. Kanellopoulos, "*Analysis of 15-Months Rain Rate Measurements at NTUA Campus*", 6th EuCAP, 26-30 March, Prague Czech Republic.

[43] J. Lemorton, L. Castanet, F. Lacoste, C. Riva, E. Matricciani, U.-C. Fiebig, M. Van de Kamp, A. Martellucci, "Development and Validation of Time-Series Synthesizers of Rain Attenuation for Ka-Band and Q/V-Band Satellite Communication Systems", *International Journal of Satellite Communications and Networking,* 2007.

[44] ESTEC Contr. 16865/03/NL/EC: "Development of Propagation Models for Telecommunication Satellite Systems", *Final Report*, 2004.

[45] L. Luini, C. Capsoni, "A Methodology to Generate Cloud Attenuation Fields from NWP Products", 6th EuCAP 2012, 26-30 March 2012, Prague, Czech Republic.

[46] M. Cheffena, L. E. Braten, T. Ekman, "On the Space-Time Variations of Rain Attenuation", *IEEE Transactions on Antennas and Propagation*, vol. 57, no.6, pp. 1771-1782, 2009.

[47] G. A. Karagiannis, A. D. Panagopoulos, J. D. Kanellopoulos, "Multi-Dimensional Rain Attenuation Stochastic Dynamic Modeling: Application to Earth-Space Site Diversity Systems", accepted in *IEEE Transactions on Antennas and Propagation.*

[48] N. Jeannin, L. Feral, H. Sauvageot, L. Castanet, F. Lacoste, "A Large-Scale Space-Time Stochastic Simulation Tool of Rain Attenuation for the Design and Optimization of Adaptive Satellite Communication Systems Operating between 10 and 50 GHz", *International Journal of Antennas and Propagation*, vol. 2012, 2012.

[49] N. Jeannin, G. Carrie, L. Castanet, F. Lacoste, "A Space Time Channel Model for the Simulation of Total Attenuation Fields", *ESA Workshop on Radiowave Propagation,* Noordwijk, Netherlands.

[50] M. Van de Kamp, "Statistical Analysis of Rain Fade Slope", *IEEE Transactions on Antennas and Propagation*, vol. 51, no. 8, pp. 1750-1759, 2003.

[51] ITU-R. P. 1623-1, "Prediction Method of Fade Dynamics on Earth-Space Paths", *International Telecommunication Union*, Geneva, 2005.

[52] B. Gremont, M. Filip, P. Gallois, S. Bate, "Comparative Analysis and Performance of Two Predictive Fade Detection Schemes for Ka-Band Fade Countermeasures", *IEEE Journal on Selected Areas in Communications*, vol. 17, no. 2, pp. 180-192,1999.

[53] C. I. Kourogiorgas, S. A. Kanellopoulos, A. D. Panagopoulos, S. N. Livieratos, G. E. Chatzarakis, J. D. Kanellopoulos, "Rain Attenuation Hitting Time Statistical Distribution: Application to Fade Mitigation Techniques of Future Satellite Communication Systems", Accepted in *1st IEEE ESTEl*, Rome, Italy, October, 2012.

[54] Paraboni, M. Buti, C. Capsoni, D. Ferraro, C. Riva, A. Martellucci, P. Gabellini, "Meteorology-Driven Optimum Control of a Multibeam Antenna in Satellite Telecommunications", *IEEE Transactions on Antennas and Propagation,* vol. 57, no. 2, 2009.

[55] D. Panagopoulos, P.-D. M. Arapoglou, G. E. Chatzarakis, J. D. Kanellopoulos, P. G. Cottis, "Long-Term Rain Attenuation Probability and Site Diversity Gain Prediction Formulas", *IEEE Transactions on Antennas and Propagation*, vol. 53, no. 7, pp. 2307-2313, 2005.

[56] V. Bosisio, C. Riva, "A Novel Method for the Statistical Prediction of Rain Attenuation in Site Diversity Systems: Theory and Comparative Testing against Experimental Data", *Internation Journal of Satellite Communications,* vol. 16, pp. 47-52, 1998.

[57] I. Kourogiorgas, A. D. Panagopoulos, J. D. Kanellopoulos, "On the Earth-Space Site Diversity Modeling: A Novel Physical-Mathematical Outage Prediction Model", *Accepted in IEEE Transactions on Antennas and Propagation.*

[58] P.-D. M. Arapoglou, A. D. Panagopoulos, P. G. Cottis, "*An Analytical Prediction Model of Time Diversity Performance for Earth-Space Fade Mitigation*", Hindawi Publishing Corporation, 2008.

[59] V. Fabbro, L. Castanet, S. Croce, C. Riva, "Characterization and Modelling of Time Diversity Statistics for Satellite Communication from 12 to 50 GHz", *Internation Journal of Satellite Communications*, vol. 27, 2009.

In: Recent Advances in Satellite Research and Development ISBN: 978-1-62417-443-8
Editors: Samuel Gardiner and Kenneth P. Olsen

Chapter 3

SATELLITE APPLICATIONS FOR VERY SHORT-RANGE WEATHER FORECASTING SYSTEMS IN SOUTHERN AFRICAN DEVELOPING COUNTRIES

Estelle de Coning[•]
South African Weather Service, Pretoria, South Africa

ABSTRACT

The strength of nowcasting and very short-range forecasting (up to 12 hours lead time) lies in the fact that it provides location-specific forecasts regarding the initiation, growth, movement and dissipation of weather phenomena. In an ideal world, weather radar systems are the most important part of nowcasting. The harsh reality, however, is that many developing countries (DC), and even more so Least Developing Countries (LDC), do not have operational radar systems at all. Furthermore, the countries which are fortunate enough to have radar systems, are struggling to maintain and sustain these powerful data sources. Nowcasting and very short-range forecasting systems for developing countries and/or least developing countries should thus be focusing on using tools such as satellite data and other observation data to address these needs.

The usefulness of Meteosat Second Generation (MSG) geostationary satellite products for nowcasting purposes goes far beyond the availability of 12 channels which can be visualized every 15 minutes. It also includes satellite-based instability indices which aide in the very short-range forecasting of convection as well as other products supplied by the Nowcasting Satellite Application Facility (SAFNWC). Satellite-based precipitation estimations find very valuable applications in the World Meteorology Organization's (WMO) Flash Flood Guidance System (FFGS) projects in various developing regions of the world. The FFGS is a hydrometeorological modeling system that predicts the potential for flash floods based on hydrological conditions that may lead to flooding in a specific basin. The Southern African Regional Flash Flood Guidance (SARFFG) project is one of the sub-regional projects of the WMO's global FFGS program. SARFFG is highly dependent on satellite-based precipitation estimations to determine whether small catchments have the potential to experience flash floods.

[•] E-mail: estelle.deconing@weathersa.co.za.

The WMO Severe Weather Forecasting Demonstration Projects (SWFDP) aim to improve forecasting and severe weather forecasting in DC and LDC by providing access to forecasting information through a website. Feedback from the participating National Meteorological and Hydrological Services in the Southern African SWFDP indicates that the SWFDP has assisted to improve warning services in many countries. At the fourth meeting of the WMO Commission for Basic Systems SWFDP Steering Group in Geneva, February 2012, it was recognized that one of the main challenges for the SWFDP was "the need for very short-range forecasting tools, to address especially the rapid onset of localized severe thunderstorms which can produce heavy precipitation and strong wind, given the absence of adequate real-time observational networks, especially weather radar coverage." It was also agreed that real-time satellite rainfall estimates have proven particularly useful in regions where radar coverage and rain gauges are sparse.

In this chapter an overview and examples will be provided of recent applications and products which were developed using geostationary satellite data to improve nowcasting and very short-range forecasting of convection, precipitation and flash floods in areas where other remote sensing tools are not readily and/or operationally available.

1. Introduction

According to the World Meteorological Organization (WMO) Public Weather Service (PWS) program, nowcasting "comprises the detailed description of the current weather along with forecasts obtained by extrapolation for a period of 0-to-6 hours ahead" (http://www.wmo.int/pages/prog/amp/pwsp/Nowcasting.htm). Very short-range forecasting extends up to twelve hours. Nowcasting and very short-range forecasting thus covers the first twelve hours of a forecast. In this time range it is important to issue warnings to the general public of hazardous weather including tropical cyclones, thunderstorms and tornados which are accompanied by flash floods, lightning and destructive wind. For the aviation industry nowcasting is useful for take-off and landing situations as well as in the *en-route* environment.

The main focus of a forecaster's task in these first few hours of a forecast is on the latest data from remote sensing tools such as radar and satellite, as well as surface observations. In developed and some developing countries, where radar systems are readily available, nowcasting relies heavily on radar data since it provides information on the size, shape, intensity, speed and direction of movement of individual storms on a continuous basis. However, the harsh reality is that many developing countries (DC), and even more so Least Developed Countries (LDC), do not have operational radar systems, and the countries which are fortunate enough to have radar systems are struggling to maintain and sustain these powerful, yet complex, data sources.

Many of the countries in Africa can be described as Developing Countries; 33 countries were classified as Least Developed Countries in 2012 (UN Secretary-General Report 2012 on the Implementation of the Istanbul Program of Action for Least Developed Countries for the Decade 2011-2020). Figure 1 provides a map depicting the location of the LDC. According to the Global Humanitarian Forum (2008), DC have the least number of ground-level, weather data observation systems, which also form the basis for efficient delivery of weather information. They also state that the African continent, which covers a fifth of the world's total land area, "has the least developed land-based weather observation system of all continents, and one that is in a deteriorating state".

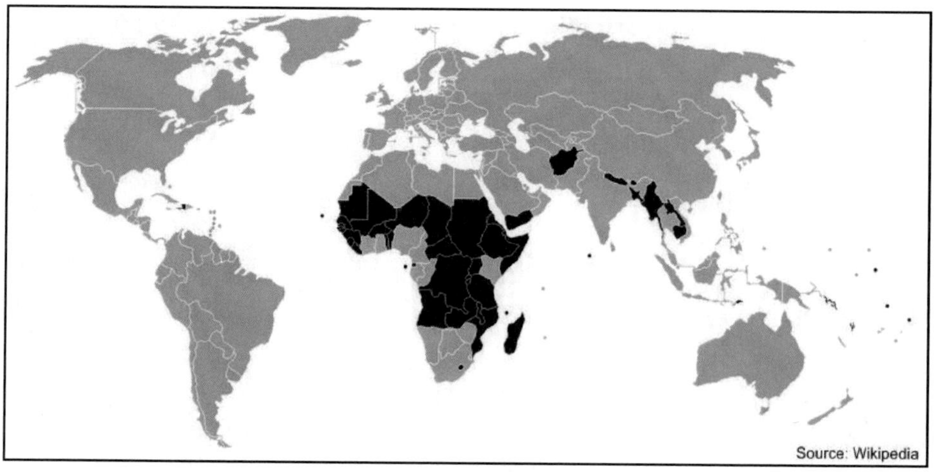

Figure 1. Map of Least Developed Countries (after, http://www.elrst.com/tag/least-developed-countries/, March 2011).

Despite the fact that weather stations, and sometimes even radar systems, exist in many DC and LDC, many of these are not operational or do not operate at all. Ideally, it is estimated that 10 000 weather stations should be operating in Africa, while only less than a quarter of this number provide observations that meet WMO requirements for standards and frequency of data.

Contrary to data availability limitations from expensive and complex weather radar systems, satellite data with excellent spatial and temporal resolution is available in most DC and LDC. This data source can supply forecasters with much needed information on various weather phenomena in real time and assist them in issuing more accurate forecasts and warnings. Addressing the need for nowcasting and very short-range forecasting systems for developing countries and/or least developed countries in Africa, implies that one will have to look beyond the use of only ground-based radar systems. The emphasis should be on using what is available, without giving up on efforts, with international assistance when possible, to build upon, and improve the coverage and sustainable operation of observational networks (e.g., surface rain gauge stations, ground-based weather radar systems etc.).

Polar orbiting satellites, such as the National Oceanic and Atmospheric Administration Polar Orbiting Environmental Satellite (NOAA/POES), EUMETSAT Meteorological Operational satellite (MetOp), Aqua, Terra, CloudSat, and the Suomi National Polar-orbiting Partnership have the advantage of high spatial resolution (~250m), but the disadvantage that temporal coverage is in the form of narrow swaths at times and places which are not always coinciding with rapidly developing weather systems. Thus, these satellites are not useful for operational forecasting purposes, although the data can be useful for research purposes and provide a critical input to the data assimilation systems for Numerical Weather Prediction (NWP). The Geostationary Operational Environmental Satellite (GOES) and Meteosat Second Generation (MSG) are located much higher above the earth surface and thus have a coarser resolution (~1 to 3 km) than the polar orbiting satellites. In contrast, they have the major advantage that images are available in near real time for the entire footprint of the satellite. Forecasters who are using satellite images can thus easily monitor and forecast the movement and changes in intensity of large and small scale features (Figure 2).

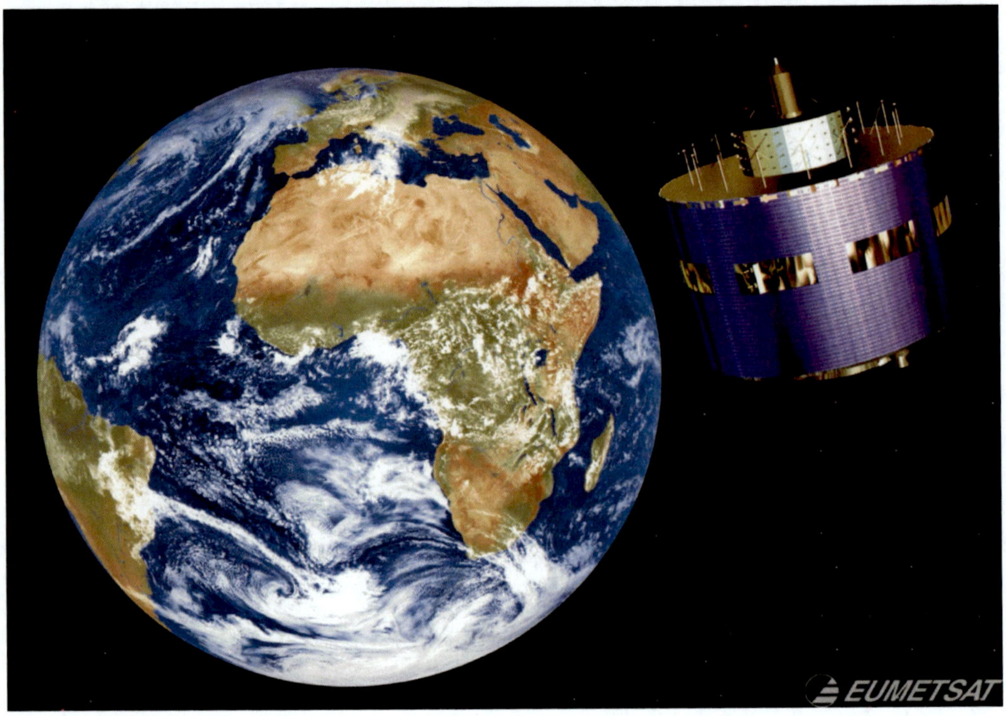

Figure 2. MSG view of Africa, copyright EUMETSAT.

In this chapter the role of the MSG geostationary satellite, the data of which is available over the entire African continent, will be highlighted through examples of various applications in products used for forecasting weather phenomena in the first few hours of a forecast. In section 2, background information will be provided on the MSG satellite. Recently developed applications of satellite data for forecasting for the first 12 hours over the southern African continent will be discussed in section 3. In the following section a major project of the WMO for developing countries will be addressed as a way to disseminate the information in a reliable fashion to DC and LDC. A summary and conclusion will be provided in section 5.

2. Meteosat Second Generation Geostationary Satellite

The European Organization for the Exploitation of Meteorological Satellites (EUMETSAT) owns four new-generation satellites. The first satellite of the series was launched on August 28, 2002. After an intensive commissioning phase, this satellite (called Meteosat-8 or MSG-1) was in full operation by January 29, 2004, and the dissemination of data commenced on a routine basis throughout Europe, Africa and the Middle East (Morgan, 2002).

MSG-2 (Meteosat Second Generation-2) is the follow-on to MSG-1 and was launched on December 21, 2005. The second in the series carries the same instruments as MSG-1, namely the Spinning Enhanced Visible and InfraRed Imager (or SEVIRI) and Geostationary Earth Radiation Budget (GERB).

Table 1. SEVIRI Imaging Channels (adapted from Morgan, 2002, MSG Brochure: Applications of MSG)

Channel name	Nominal central wavelength (µm)	Applications
VIS 0.6	0.635	Cloud detection, cloud tracking, scene identification, monitoring of land surfaces and aerosols, and in combination to generate vegetation indices.
VIS 0.8	0.81	Cloud detection, cloud tracking, scene identification, monitoring of land surfaces and aerosols, and in combination to generate vegetation indices.
NIR 1.6	1.64	Discriminate between snow and clouds, as well as between ice and water clouds, aerosol information.
IR 3.9	3.92	Detection of low clouds and fog at night, measurement of land and sea temperatures at night, the detection of forest fires.
WV 6.2	6.26	Measurements of mid-atmosphere water vapor, tracers for atmospheric winds, height assignment for semi-transparent clouds.
WV 7.3	7.35	Measurements of mid-atmosphere water vapor, tracers for atmospheric winds, height assignment for semi-transparent clouds.
IR 8.7	8.7	Provide quantitative information on thin cirrus clouds, discrimination between ice and water clouds.
IR 9.7	9.66	Ozone concentration in the lower stratosphere, to monitor total ozone, assess diurnal variability, tracking ozone patterns as an indicator of wind fields at that level.
IR 10.8	10.8	Temperature of clouds and the surface, help to reduce atmospheric effects when measuring surface and cloud top temperatures, cloud tracking for atmospheric wind and for estimates of atmospheric instability.
IR 12.0	12.0	Temperature of clouds and the surface, help to reduce atmospheric effects when measuring surface and cloud top temperatures, cloud tracking for atmospheric wind and for estimates of atmospheric instability.
IR 13.4	13.4	Estimation of atmospheric instability, temperature information on the lower troposphere.

The satellite was renamed Meteosat-9 when it became operational in June 2006. MSG-2 has twelve channels to use individually or in combination for various purposes. Image pixels are sampled every 15 minutes at intervals of 3 km over the entire earth disc within view for eleven of the twelve channels.The High Resolution Visible (HRV) channel has a sampling distance of just 1 km, with the east-west scan limited to half of the full earth disc. It is planned that MSG-2 will be operational until 2018, where after there will be a follow-on series in geostationary orbit called Meteosat Third Generation (MTG). Table 1 gives a summary of the different channels as well as their applications.

EUMETSAT is involved in continuous training events in Europe, Africa and South America through which the use of the MSG channels are explained and demonstrated. Individual channels, as well as channel differences, can be combined in so-called Red-Green-Blue (RGB) combinations to highlight specific cloud characteristics such as cloud depth, phase or height of the cloud.

Table 2. Commonly used RGBs and their applications (adapted from MSG Interpretation Guide)

RGB name	RGB composite	Application
Dust /Ash	Channels 10-09, 09-07, 09	Dust, thickness and phase of clouds, contrails, fog, ash, SO2, low-level humidity
Air mass	Channels 05-06, 08-09, 05	Severe cyclones, jets, Potential Vorticity analysis
Night microphysical	Channels 10-09, 09-04, 09	Clouds, fog, contrails, fires
Day Microphysical	Channels 02, 04r, 09	Clouds, convection, snow, fog, fires
Convective storms	Channels 05-06, 04-09, 03-01	Severe convection
Day Solar	Channels 02, 03, 04r	Snow, fog
Day Natural	Channels 03, 02, 01	Vegetation, snow, smoke, dust, fog

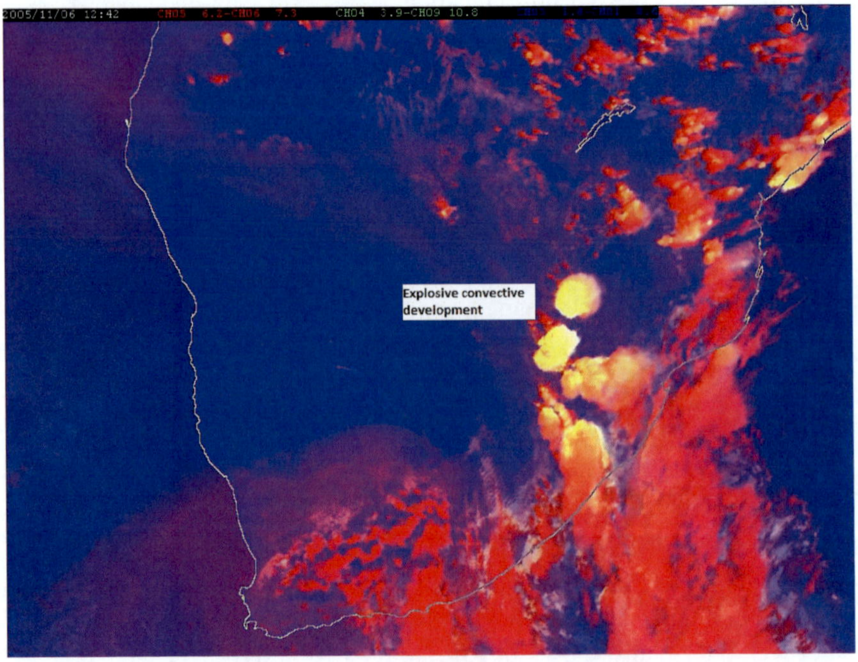

Figure 3. Convective storms RGB (Channels 05-06, 04-09, 03-01) on November 6, 2006 at 12:30 UTC, annotation courtesy of Jochen Kerkmann, copyright EUMETSAT.

For the purpose of simplicity and standardization, a few of the most common RGB combinations have been identified (Table 2). A full list of all the products and their interpretation can be found in the MSG Interpretation Guide available at: http://oiswww.eumetsat.org/WEBOPS/msg_interpretation/index.php.

An example of how one of these RGB combinations can be used to indicate areas of convection and even severe convection is shown in Figure 3. Using this Convective storms RGB, the yellow colors indicate where tall, cold clouds can be found containing strong updrafts and small ice crystals. In this example, severe wind damage as well as heavy downpours, were recorded in the area where the bright yellow colors were found (Koenig and de Coning, 2006).

Aside from the efforts to conduct hands-on training events for forecasters in all countries which receive MSG data, a web portal was also developed where case studies of numerous weather events from several countries are presented and analyzed using mainly MSG data. The satellite data is also supplemented with radar and/or numerical weather prediction model data, where available and useful. This website (http://www.eumetsat.int/Home/Main /Image_Gallery/Case_Studies/index.htm?l=en) is a very practical way of teaching scientists and forecasters about the use of MSG imagery.

MSG makes it possible to provide images during the day and/or night time of clouds and cloud systems. It is also possible to get information on the internal processes and microphysical state of different clouds and obtain insight into related environmental parameters (e.g., cloud droplet size can be inferred, and surface fog may be detected even at night and vegetation growth monitored). In addition to the purely visual interpretation of the MSG channels, the ultimate usefulness comes from digital products extracted from the data.

The Satellite Application Facilities (SAF) in Europe have developed products for very specific applications (http://www.eumetsat.int/Home/Main/Satellites/GroundNetwork/ ApplicationGroundSegment/SAFs/WhatisaSAF/index.html). The SAF are dedicated centers of excellence for processing satellite data and derive value-added products. Specialist expertise from the European Union Member States is used to develop products which are disseminated to all the member countries. Currently, there are eight SAF providing operational products and services:

- SAF on Nowcasting and Very Short Range Forecasting
- SAF on Ocean and Sea Ice
- SAF on Climate Monitoring
- SAF on Numerical Weather Prediction
- SAF on Land Surface Analysis
- SAF on Ozone and Atmospheric Chemistry Monitoring
- SAF on Radio-Occultation Meteorology
- SAF on Support to Operational Hydrology and Water Management.

The SAF on Nowcasting and Very Short-Range Forecasting offers a long list of products which are aimed at the nowcasting time range, which include clouds, precipitation and convection, clear air as well as wind products. The software to create the different products can be downloaded subject to a license agreement. Several European countries are making use of the Nowcasting SAF products on an operational basis, and in South Africa research experiments have been conducted using this software.

A wide range of satellite-derived products have been derived centrally within the MSG Meteorological Products Extraction Facility (MPEF) located at the EUMETSAT Headquarters in Darmstadt, Germany (Morgan, 2002). The first step in the MPEF processing is an automatic scene analysis, using multispectral thresholding techniques, of each image in near real time, to decide into which class each pixel of the image falls (e.g., sea or land, low or high clouds, etc.). This comprehensive analysis of each image forms the basis for further processing to calculate other MPEF products. These products are distributed on the Global Telecommunications System (GTS) of the WMO, as well as via the EUMETSAT dissemination service, called EUMETcast (Morgan, 2002). A list of some of the MPEF products is shown in Table 3; new products are added according to user needs.

Table 3. List of MPEF products supplied by EUMETSAT (http://www.eumetsat.int)

PRODUCT	PURPOSE
AMV (Atmospheric Motion Vectors)	Most important product for Numerical Weather Prediction (NWP). Applications: information on atmospheric dynamics not available from other sources.
CSR (Clear Sky Radiances)	CSR are the mean radiances for cloud-free pixels. Applications: useful for operational NWP.
GII (Global Instability Indices)	GII are airmass parameters which act as stability indicators of the atmosphere. Applications: used to complement upper-air sounding data on a real-time basis.
TH (Tropospheric Humidity)	TH indicates the layer mean humidities between 500 and 200 hPa and between 850 and 300 hPa, respectively.
CLA (Cloud Analysis)	CLA is based directly on the scene analysis. Applications: information on cloud cover, cloud top temperature / pressure / height, cloud type and phase.
CTH (Cloud Top Height)	CTH provides the height of the highest cloud. Applications: aviation meteorology.
CDS (Climate Data Set)	CDS provides information on the scene classes in each segment. Applications: climate studies.
TOZ (Total Ozone)	The TOZ product provides information on ozone. Applications: information on high-level wind can also be assimilated directly into NWP models.
IDS (ISCCP Data Set)	The IDS is the EUMETSAT/MSG contribution to the International Satellite Cloud Climatology Project (ISCCP).
HPI (High Resolution Precipitation Index)	The HPI is the EUMETSAT/MSG contribution to the Global Precipitation Climatology Project (GPCP). Applications: tropical convective rainfall.

3. Recent Development in Satellite Applications for Nowcasting and Very Short-Range Forecasting

A working group has been established to discuss and coordinate research in the field of convection. The aims and activities of this Convection Working Group are available at http://ww.convection-wg.org. A so-called "Best Practice document" was compiled by the members of this group to describe the established convection nowcasting methods. What follows in this paragraph summarizes some of the relevant practices from this document.

Recent developments in the use of satellite data for nowcasting and very short-range forecasting of convection can be divided into three stages: the pre-convective stage, the stage where convection is initiated and then the mature convection stage. Although satellite data cannot be seen or treated as the only data set to provide information for any of these stages, very useful products have been developed in the last few years to assist a forecaster in the process of nowcasting and very short-range forecasting. Satellite data and products can serve as a near real time input together with observations, numerical weather prediction products as well as data from other remote sensing tools.

3.1. The Pre-Convective Stage

In the pre-convective stage, satellite and numerical weather prediction fields can be used in combination to provide information on atmospheric conditions and how they can develop in the next six-to-nine hours. Petersen et al. (2010) has developed a "nearcast" product based on the differences of equivalent potential temperatures as well as moisture in the vertical profile of the atmosphere. Georgiev (2003), Rabin et al. (2004) and Schmetz et al. (2005) have made use of satellite-derived wind vectors and water vapor to analyze moisture boundaries and their movements. These products can be utilized in developing countries but are not operationally available over the African continent yet.

3.1.1. Satellite Based Instability Indices

a) Global Instability Indices (GII)

An important method used by forecasters to assess the possibility of thunderstorm development early in the morning, is by making use of information of the vertical profile of the atmosphere. Using the temperature and dew point temperature information at various levels of the atmosphere, indications of the instability of the atmosphere can be obtained; if the atmosphere is unstable, the likelihood of thunderstorms is greater. Various so-called instability indices can be derived to describe the atmospheric conditions (Peppler, 1988). The indices values can, however, vary a lot with time and space because of frequent changes in wind and low-level moisture in storm environments. Traditionally, these vertical profiles are obtained from radiosonde profiles at designated weather stations. Unfortunately, the instruments used in radiosonde ascents are relatively expensive to many DC and LDC, and many of these observations are needed to have accurate information on the atmospheric conditions (Wagner et al., 2008) at various locations.

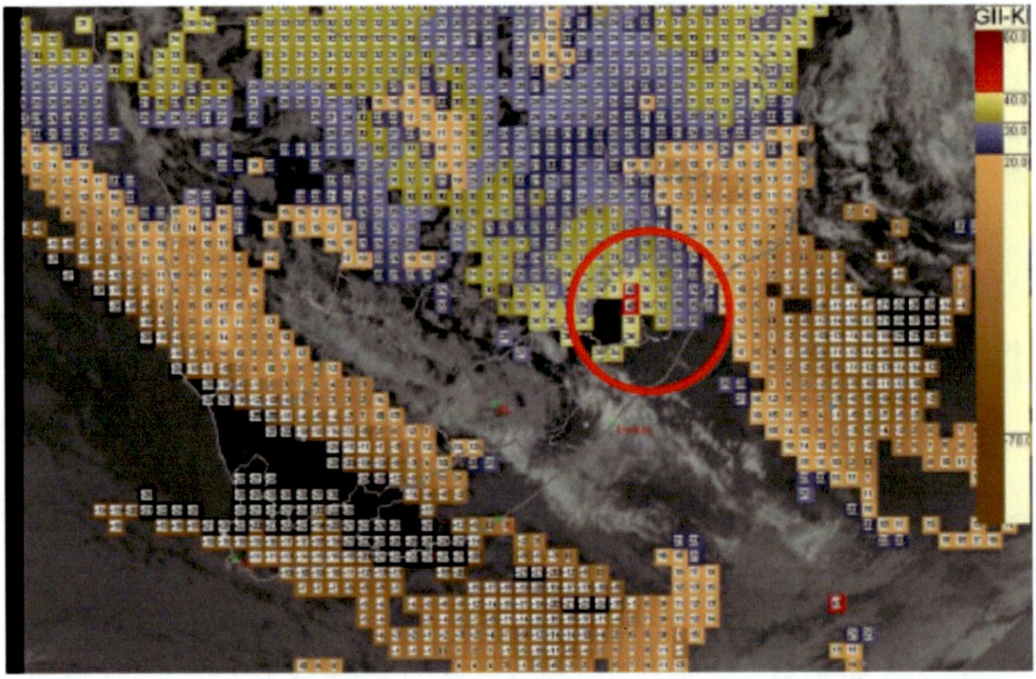

a.

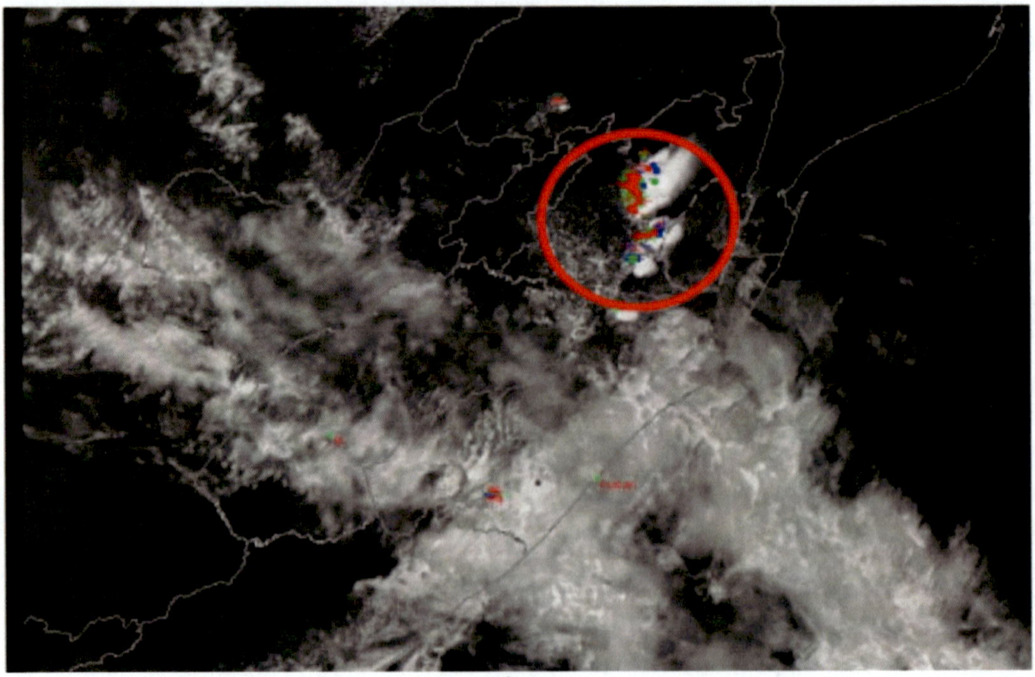

b.

Figure 4. a) Satellite-based K-Index over South Africa, for October 26, 2006, 08:00 UTC. Values of increasing instability are shown in blue to yellow to red. b) Image of the MSG high resolution VIS channel overlaid with lightning for October 26, 2006, 10:30 UTC (after Koenig and de Coning, 2009).

Satellite-based temperature and moisture retrievals of instability and air mass parameters have been operationally in use since 1988, utilizing the GOES VISSR Atmospheric Sounder (VAS) instrument and later also the GOES Sounder (Hayden, 1988; Dostalek and Schmit, 2001; Schmit et al., 2002). One of the MPEF products listed in Table 2 is the Global Instability Indices (GII) - this is a suite of instability indicators based on information from the MSG satellite. The methodology uses six channels from the MSG satellite in combination with information from a numerical weather prediction model to assess the air stability in pre-convective (i.e., usually still cloud-free) conditions. The calculation of the GII makes use of numerical weather prediction (NWP) input from the European Center for Medium-range Weather Forecasts (ECMWF). The most commonly used indicators calculated for the GII are the K Index, Lifted Index and a moisture indication in the form of Total Column Precipitable Water. The methodology for this calculation is described in Koenig (2002), Koenig (2007) and Koenig and de Coning (2009).

The greatest advantage of these fields is the added capability of the nearly continuous monitoring of the instability fields guaranteed by the MSG 15-minute repeat cycle. This provides forecasters with information on the atmospheric conditions with improved temporal as well as spatial resolution compared to soundings done twice daily at only a limited number of radiosonde stations. Satellite-based instability indices, similar to radiosonde information, could help a forecaster to focus their attention on a particular region where convection is likely. It is still necessary to monitor the situation in real time and a complete picture of the thunderstorm possibilities can only be obtained when instability indices are interpreted in combination with other triggering and/or lifting mechanisms (Koenig and de Coning, 2009).

These authors showed that the GII over Europe compares well with the indices derived from radiosonde information. The early morning values of the K Index and Lifted Index of the GII over South Africa were also compared to the occurrence of lightning later in the same day, and it was clear that the satellite-based instability indices could accurately determine the location of convective development with a lead time of three-to-nine hours.

Figure 4 shows an example of the K Index on October 26, 2006 at 08:00 UTC. The highest values of the K Index (>41°C) were over the eastern part of South Africa, extending into Mozambique. At 10:30 UTC, lightning occurred in the identified area. The lightning strokes on the display are either "-" or "+" signs, indicating the polarity of the stroke. Very thick "+" or "-" symbols indicate lightning strokes with an intensity in excess of 50 kAmpere. The red symbols are all strokes reported in the prior 5 minutes, the blue symbols are those reported 5 to 10 minutes ago while the green symbols are those reported between 10 and 15 minutes ago.

b) Regional Instability Indices (RII) and Combined Instability Index (CII)

The South African Weather Service (SAWS) runs a local version of the UK Meteorological Office Unified Model (UM) twice daily to provide hourly numerical forecasts of atmospheric conditions for up to 48 hours ahead. The model domain covers the area between 0.48° N and 44°S, and 10°W and 56°E (i.e., southern Africa). Currently, the East-West resolution of the model is 0.11° and the North/South resolution is 0.1112° - which translates to more or less 12km. In September 2007 a local version of the GII code was installed at SAWS using the local version of the Unified Model instead of the ECMWF model. This southern African version of the GII is called the Regional Instability Indices

(RII). The RII product does not cover the entire MSG footprint, but only the domain of the Unified Model run on South African servers (i.e., southern Africa) (Figure 5).

De Coning et al. (2011) added Total Totals to the suite of satellite derived instability indices and used a Modified K Index, as well as a Modified Total Totals (Charba, 1977), to compensate for the areas with higher elevations in South Africa. An effort was also made to combine the standard ingredients for convection (instability, moisture and lift) into one probability map (expressed in percentages) for convective activity. The combination involved the cumulative frequencies of lightning related to each of the four instability indices from the RII (Modified K index, Lifted Index, Modified Total Totals), the Total Column Precipitable Water as a moisture parameter and topography, which acts as a lifting mechanism.

Together with the cumulative frequency values for all meaningful values of each RII, the Hanssen-Kuipers discriminant (Wilks, 2005) value for that value of the RII was taken into account as a weighting factor for the combination. The Combined Instability Index (CII) was then defined as the combination of the three instability indices from RII, the PW from the RII and height above sea level.

Figure 5. Regional Instability Index (Lifted Index) on August 20, 2012 at 06:30 UTC showing the domain of the Unified Model. Picture courtesy of the South African Weather Service, SUMO display system.

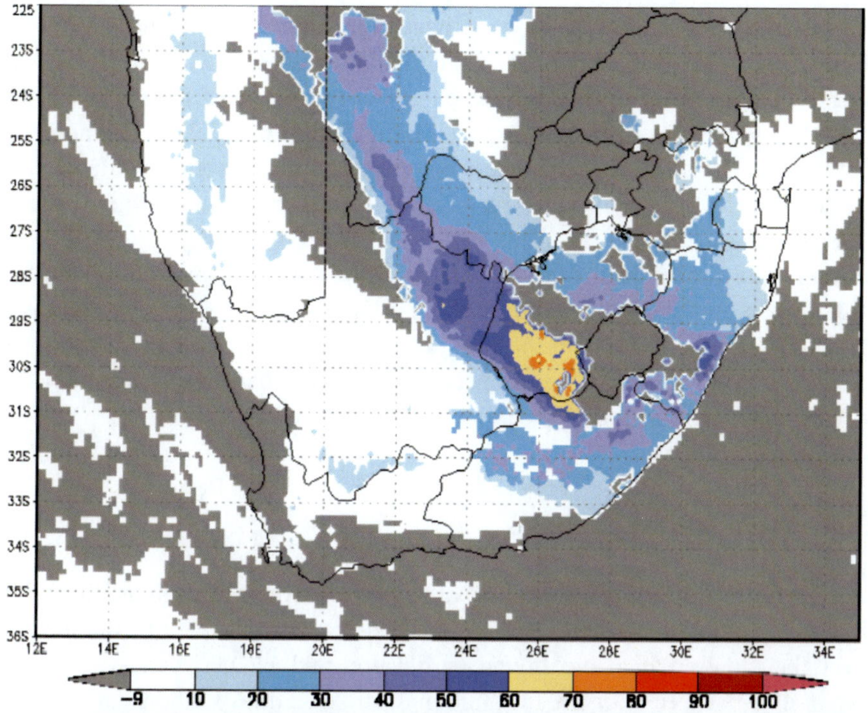

a.

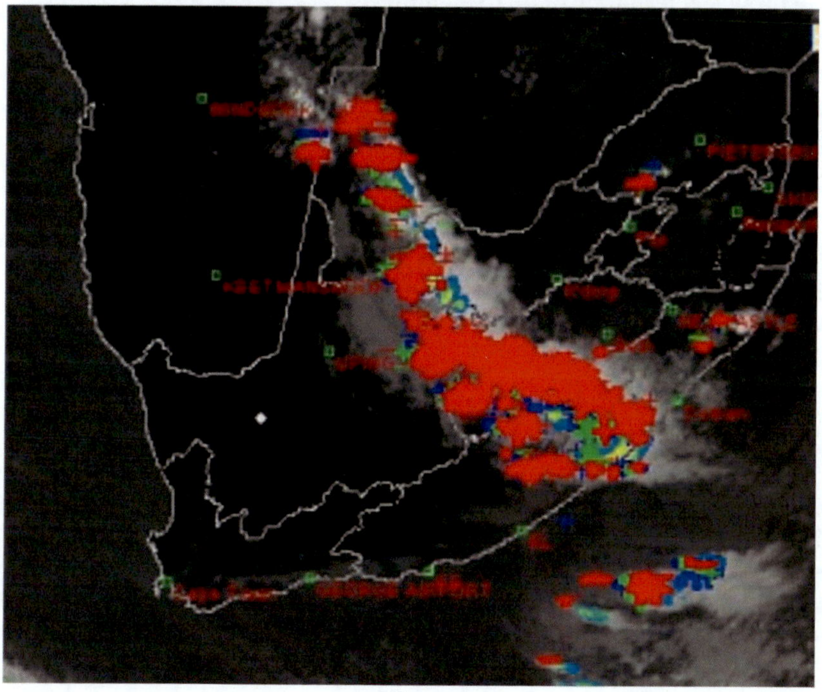

b.

Figure 6. a) CII time average between 06:00 and 09:00 UTC on January 31, 2010. The grey color indicates the cloud cover, where the satellite based product cannot be calculated; and b) MSG IR108 overlaid with lightning at 15:00 UTC, after de Coning et al., 2011.

The CII early in the morning was shown to have high statistical evaluation scores which also outperformed the statistical scores of the individual parameters when evaluated against the occurrence of lightning later in the day. The biggest advantage of this parameter is the fact that it can provide a lead time for the occurrence of convection with reasonable accuracy for the entire southern Africa – also in areas where there are no radar systems available. The CII is also available every 15 minutes, with every new MSG image. An example of the CII is shown in Figure 6.

On this summer day, the 3-hour time averaged field of the CII from 06:00 to 09:00 UTC showed that there is a probability for convection in a band stretching from the northwest to the southeast across South Africa, with the best probability (>60%) over the central parts of South Africa. At 15:00 UTC lightning occurred in this same band with large amounts of lightning over the area where the best probability for convection was.

3.2. The Convection Initiation Stage

This stage involves the growing and intensification of existing cumulus clouds during which the first precipitation is detected. Various methods have been developed in the last few years to detect the clouds which are bound to develop and intensify and cause the first detection of a 35 dBz radar echo or the first lightning stroke. These methods have shown the possibility to provide a lead time of up to 90 minutes.

A cloud identification method developed by Berendes et al. (2008) has been used by other algorithms for convective weather forecasting. Mecikalski et al. (2007) showed that this methodology could be very useful for the identification of over-shooting tops and this has applications in the aviation industry.

Mecikalski and Bedka (2006) used several channels from MSG (so-called "interest fields") to identify the precursor signals of convective initiation. An object-tracking approach was incorporated in this scheme in 2010 (Mecikalski et al., 2010) with positive validation results. The research in this field is ongoing, but has not yet been incorporated in nowcasting systems in DC in Africa.

3.3. Mature Stage

In the mature stage, clouds are present with a tall structure and possibly strong updrafts and ice crystals. These types of convective elements can be the cause of severe weather at the surface such as hail and/or tornadoes. Visible and infrared channels of satellites can be used to learn more about the intensity, severity, height, lifetime and precipitation rate of a mature convective storm.

Zinner et al. (2008) developed an algorithm to detect and track cumulonimbus clouds using the high resolution visible, water vapor and infrared channels of MSG, called Cb-TRAM. The Cb-TRAM algorithm has been tested against the occurrence of lightning over the South African region (Forster et al., 2010) for a number of case studies and proved to have very valuable information on the different stages of convective storms over this region. This methodology is also finding application in the aviation industry.

Setvak et al. (2010) provides the background and characteristics of cold-U/V and ring shaped storms which proves to relate very well to severe weather on the ground. This feature is easily detected using color-enhanced infrared imagery. De Coning et al. (2009) showed how a cold U-shaped storm over South Africa and Swaziland could be associated with severe storms over the same area, Figure 7.

Rosenfeld et al. (2008) used the effective radius and thermodynamic characteristics of clouds, retrieved from satellites, to identify vertical profiles which are associated with severe and non-severe convective clouds. Their methodology has been used to some extent in research on severe weather events in South Africa, but is not operationally available yet.

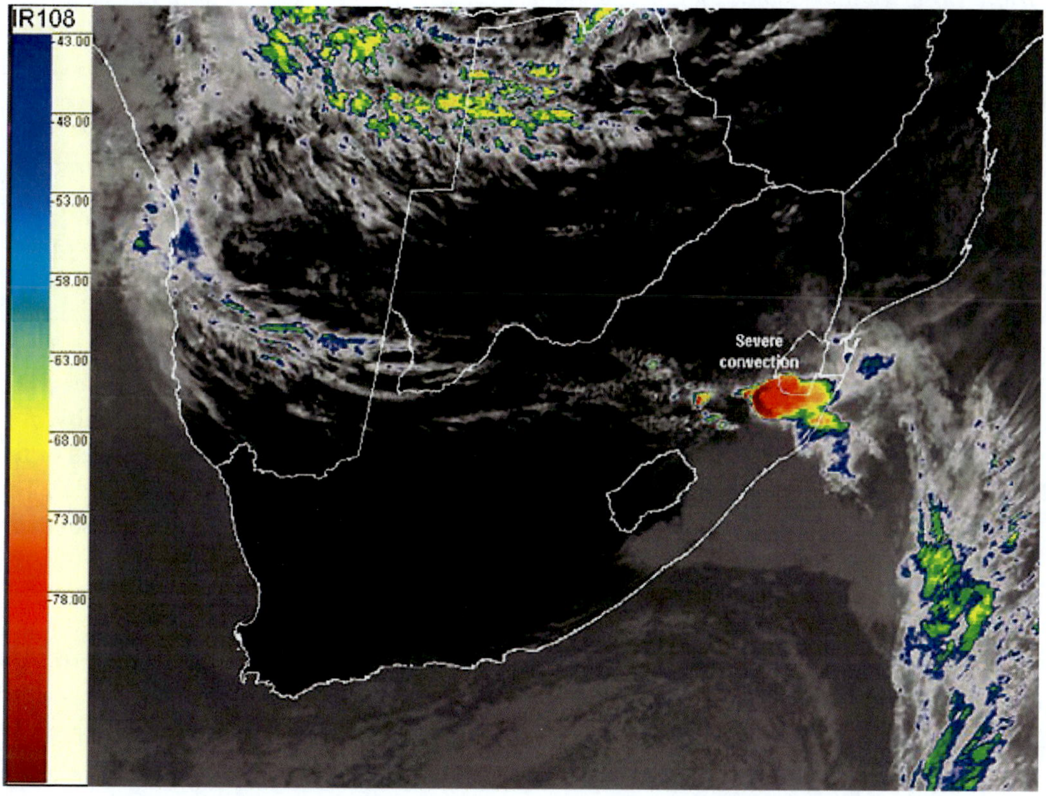

Figure 7. Severe convection over eastern South Africa indicated by a U-shaped storm in color-enhanced IR108 on November 27, 2009, 10:00 UTC. Satellite image courtesy of EUMETSAT.

3.4. Satellite Based Precipitation Estimation

3.4.1. Hydroestimator for Southern Africa

Rainfall has a high temporal and spatial variability and thus a comprehensive estimation of precipitation remains one of the most difficult observational challenges of meteorology, especially during convective rainfall events. Although rain gauges provide a direct measurement of rainfall, rain gauge networks will always be too coarse to resolve the fine-scale features of a rain field. Rain gauges are also unevenly distributed and, most importantly, they provide point source data rather than a representation of a spatial domain (Kondragunta, 2007).

Radar rainfall can be used to provide an indirect measurement of rainfall, but then the radar systems need to cover the entire area of interest, be well calibrated and have appropriate radar rainfall relationships according to the type of precipitation. For most DC, and even more so for LDC, radar systems will remain too expensive and difficult to maintain and thus will not be a feasible option for this purpose.

Satellite-based estimates of rainfall are not as accurate as gauges or radar rainfall, but has the advantage of high temporal resolution and spatial coverage, even over oceans, in mountainous regions and sparsely populated areas. In areas where there are very few rain gauges and no radar systems, satellite-derived rainfall can be a "critical tool for identifying hazards from smaller-scale rainfall and flood events." (STAR Satellite Rainfall estimates, http://www.star.nesdis.noaa.gov/smcd/emb/ff/index.php). Satellite-based Precipitation Estimators (SPE) should not be considered as a replacement for radar rainfall estimates and/or gauges, but as a complement to these fields, if at all possible (Scofield and Kuligowski, 2003).

The National Environmental Satellite, Data and Information Service (NESDIS) developed an SPE algorithm for high-intensity rainfall called the Autoestimator (AE). The original AE, developed by Vicente et al. (1998), used the 10.7 μm brightness temperatures from the NOAA satellite and related that to radar estimated rainfall. A follow-on version of the AE, called the Hydroestimator (HE) was developed which could be used outside of regions of radar coverage without compromising accuracy too much. Due to its independence of radar, the HE was found to be well suited to southern Africa. A local version of the Hydroestimator was installed at the SAWS in 2007 and has been running operationally ever since. Input files for the local HE consist of the IR10.8 channel brightness temperatures of the MSG satellite and model output fields of the SAWS local version of the Unified Model. The HE is available in the same domain as the local version of the Unified model (i.e., southern Africa). There are, of course, more accurate precipitation algorithms available, but the requirements to have a real time and simple product available which does not need input from radar systems, make the HE a suitable choice for DC and LDC in the region.

Accumulation products of the HE are created for 1, 3, 6 and 24 hours, as well as 10 days and 30 days, which can give an indication of where slow moving storms can cause local flooding. The Thunderstorm Identification Tracking Analysis and Nowcasting (TITAN) algorithm (Dixon and Wiener, 1993) was applied to the 10 mm per hour rainfall core of the HE. These tracks then give an indication regarding the direction of movement of the heavier rainfall areas for the next 60 minutes and whether the rainfall core is growing or decaying.

An example of the real-time display of the Hydroestimator is shown in Figure 8a and in Figure 8b an example is shown of the HE tracking of 10 mm per hour cores. Both these products are available in real time to forecasters in southern Africa.

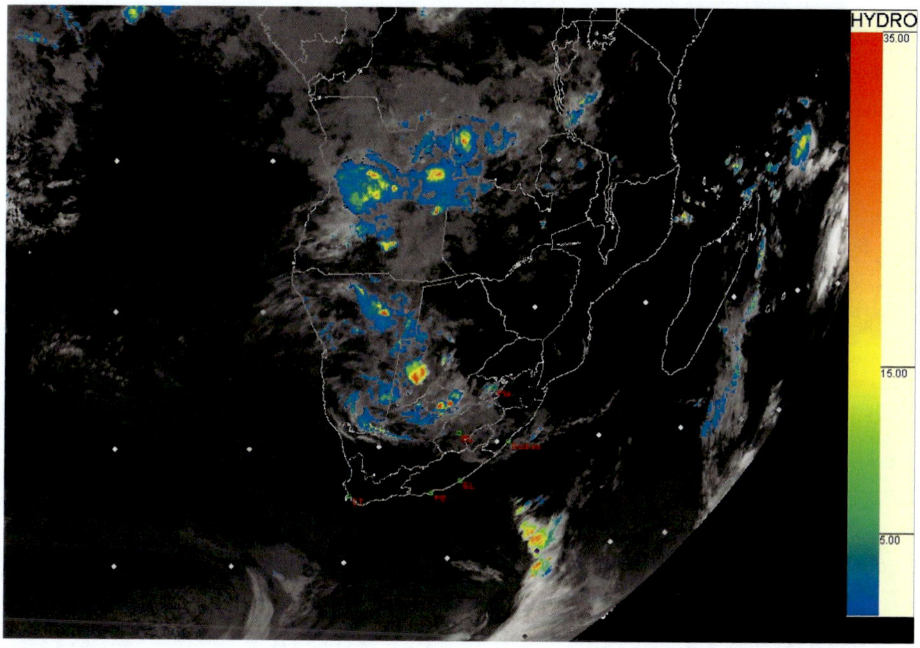

Figure 8a. Hydroestimator overlaid on MSG IR10.8 on February 9, 2012 at 05:00 UTC. Rain rate (mm) per hour indicated in the color scale on the right. Picture courtesy of the South African Weather Service, SUMO display system.

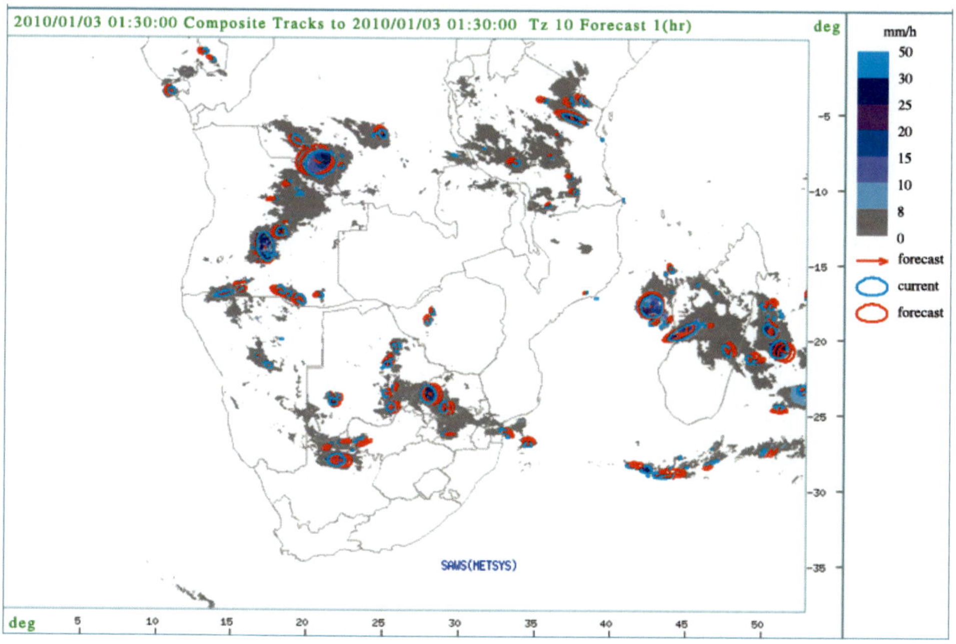

Figure 8b. Hydroestimator tracking of 10 mm per hour rainfall cores on January 3, 2010. Blue circles indicate the current position of the rainfall core and red circles predict the 1-hour predicted position of the rainfall core.

3.4.2. Satellite and Model Blended Product for Rainfall Estimation

An analysis of rainfall patterns in South Africa (Kruger, 2007) showed that summer rainfall occurs mainly over the interior of the country in the form of convective thunderstorms. In January to March, tropical rain showers are common with less heat-generated convection. In the winter months, the southwestern parts and coastal regions receive rainfall mainly as a result of cold fronts and high pressure systems ridging along the coastline. To capture the different types of rainfall, it would be necessary to include stratiform as well as convective precipitation events, in the daily rainfall data set.

The HE depends on one channel (IR10.8) and uses cloud top temperature as a first estimate of precipitation. Validation studies by analysts at the *National Environmental Satellite, Data, and Information Service* Satellite Analysis Branch (NESDIS/SAB) who work with the HE on an operational, real-time basis (Kuligowski, 2009, personal communication) conclude that the Hydroestimator is known to perform best for convective events while stratiform rainfall events are sometimes missed or under/overestimated. It was also noted that the convective events are often overestimated in terms of intensity.

A recent study (de Coning and Poolman, 2011) attempted to address these shortcomings of the HE by applying a bias correction to the HE based on the 24-hour rainfall totals of rain gauges over South Africa. A two-year period (2008 to 2009) was used to compare the rain gauge totals to the 24-hour totals of the Hydroestimator. A different bias correction was calculated for the convective season (November to April) and for the months when stratiform precipitation dominates, due to the varying performance of the HE in the different seasons.

The authors then made use of the stratiform rainfall field of the local version of the Unified Model, which was also bias corrected, and combined this with the HE in order to ensure a more complete rainfall field, including the precipitation events associated with the passage of cold fronts along the coast lines of the country (which are normally handled well by the model) as well as the convective thunderstorm events over the interior (for which the HE tends to perform better).

When the combined rainfall field was compared to rain gauge data, it was found that it proved to be better than the HE, especially in the winter months when the HE would have missed the stratiform rainfall events along the coast lines. An example of the combined precipitation product is shown in Figure 9. On this day in the winter season the stratiform precipitation field from the UM (a) underestimated the rainfall over the interior and captured some of the rain along the east coast while the HE (b) was over estimating the rainfall over the southwestern interior of South Africa and missed the stratiform precipitation which occurred along the eastern coastline. Using the combined product (c), the daily total rainfall was found to be closer to the amount of rain which was reported by the gauges (d) and included the rain along the east coast.

The frequency and extent of flooding events in South Africa necessitates attention as well as action in order to save lives and safeguard property. The SAWS and the National Disaster Management Center (NDMC) are involved in a national project for the development and implementation of a flash flood warning system in flash flood-prone regions, called the South African Flash Flood Guidance, SAFFG (Poolman, 2012). The SAFFG is based on the Flash Flood Guidance System (FFGS) which is the intellectual property of the Hydrologic Research Center (HRC), a non-profit public-benefit corporation based in San Diego, USA (Georgakakos, 2004 and 2006). The South African version of the FFGS became operational in October 2010.

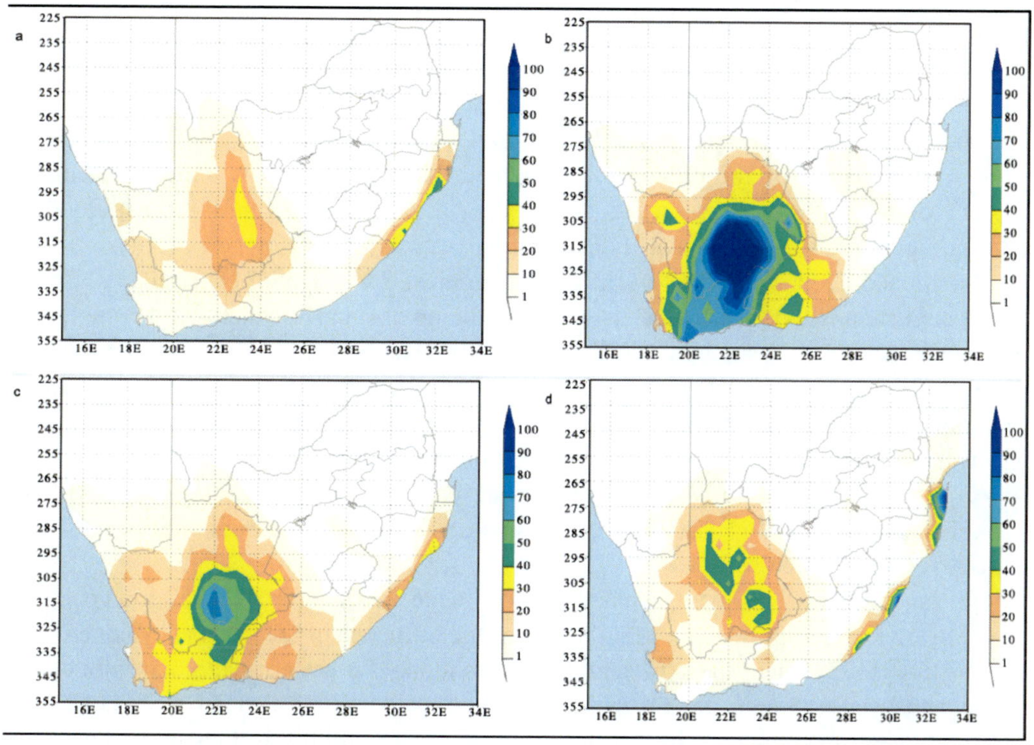

Figure 9 a) Unified Model stratiform precipitation total for 24 hours, b) Hydroestimator precipitation total for 24 hours, c) the combined product from HE and UM for 24 hours and d) the total rainfall as measured by the rain gauges for June 18, 2008 (after de Coning and Poolman, 2011).

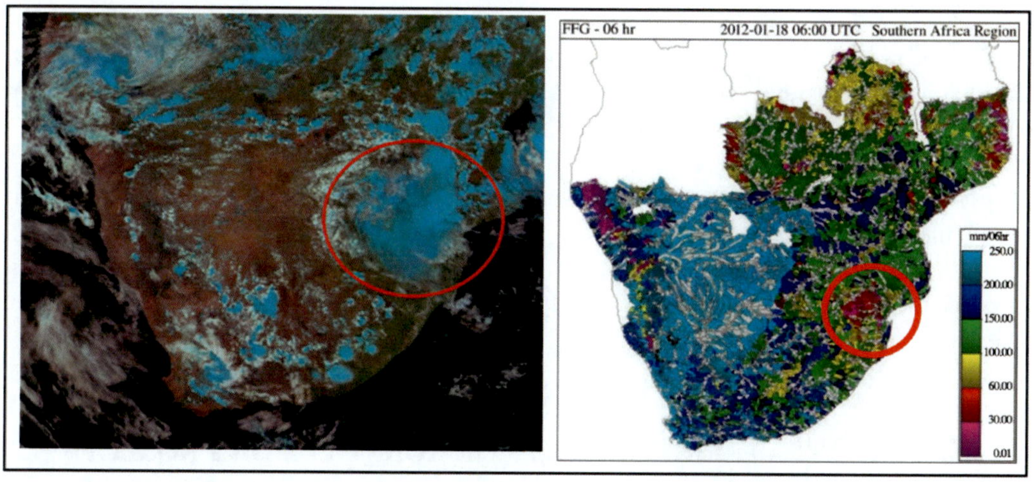

Figure 10 MSG Day Natural RGB product (left) and the SARFFG on January 18, 2012 at 06:00 UTC (right). Satellite image courtesy of EUMETSAT.

The FFGS system is a hydro-meteorological modeling system. It combines real-time meteorological information (rainfall estimations from weather radar, satellite and rain gauges) with hydrological modeling (including soil moisture conditions) and calculates a flash flood potential. This flash flood potential is calculated and graphically depicted for numerous small river basins (on average 50 km^2) in five flash flood prone regions over South Africa. In areas where there is poor or no radar coverage the system is configured for larger (200km^2) basins, and then the system uses only satellite rainfall estimates and rain fall measurements by gauges. Information of the previous 1, 3, 6 and 24 hour periods is used to calculate the amount of rainfall which is needed to fill the basin and thus cause a flash flood. This information is available and updated on an hourly basis.

In collaboration with the World Meteorological Organization a similar flash flood guidance system (called the SADC SARFFG) is being developed which will be implemented over seven countries in southern Africa during 2012.The SADC SARFFG system will cover the areas in South Africa where there are no radars and will also include six other countries (without any operational radar systems). Due to the dependence on satellite data only, the basins in the SADC SARFFG will be larger - 200km^2.

Currently, the HE hourly accumulations are one of the input fields to the SAFFG and SARFFG. This system is also web based and thus only requires internet access and no expensive visualization software.

In January 2012, tropical cyclone Dando made landfall on the eastern coastline of southern Africa. In Figure 10 an MSG image of the influence of the tropical cyclone is shown and the SARFFG, where the pink and red colors along the eastern parts of southern Africa indicate that only ~30mm of rain in a 6-hour period is needed in order to produce a flash flood. This information could be utilized by forecasters to issue warnings in the areas of risk during this heavy rainfall event.

4. Dissemination of Information in DC and LDC

Various graphical display systems to integrate satellite, radar, lightning, observations and numerical weather prediction data are available around the world. Although these systems are all highly efficient, they are also costly and require regular maintenance and upgrades. In DC and LDC it is not always practical or affordable to acquire such systems since technical support staff is not always available or equipped to deal with the needed installments and improvements. While sophisticated display systems have obvious advantages, a more simple way to disseminate and display data is required in DC and LDC. Utilizing the internet and EUMETcast, is an affordable and fairly easy way to get information to National Meteorological Centers across the continent.

Severe Weather Forecast Demonstration Projects

The WMO's Commission on Basic Systems (CBS) Severe Weather Forecasting Demonstration Projects (SWFDP) aim specifically to improve severe weather forecasting in DC as well as in LDC. Currently, there are three active SWFDP around the world - in southern Africa, the South Pacific, eastern Africa, and two under development in Southeast Asia and the Bay of Bengal (South Asia). Figure 11 gives a graphical depiction of these SWFDP showing that it covers the LDC of the world.

The goals of the SWFDP broadly include the following:

- "To improve the technical ability of weather services,
- To improve the lead time of warnings,
- To improve communication between global, regional and National Meteorological Centers (NMCs),
- To improve interaction of NMCs with disaster management authorities before and during severe weather events, and
- To identify other gaps for improvement" (Poolman et al., 2008).

Feedback from the participating National Meteorological Centers in the southern African SWFDP stated that this method of data dissemination improved warning services in many countries and contributed to the improvement of relations between National Meteorological and Hydrological Services (NMHS) and disaster management authorities. The SWFDP has contributed significantly to the end-to-end process of warning dissemination.

Although these SWFDP projects were initially designed to focus mainly on numerical weather prediction output and synoptic scale events, it is becoming clear that more than model output is needed. At the fourth meeting of the CBS-SWFDP Steering Group (CBS-DPFS related meetings, 2012), it was recognized that one of the main challenges for the SWFDP is "the need for very short-range forecasting tools, to address especially the rapid onset of localized severe thunderstorms which can produce heavy precipitation and strong wind, given the absence of adequate real-time observational networks, especially weather radar coverage". Satellite-based products were recognized as particularly useful in regions where very few rain gauges exist and little or no radar coverage is available.

The southern African SWFDP is available at the Regional Specialized Meteorological Center web portal at: http://rsmc.weathersa.co.za/RSMC/login.php for users in the NMHS of the region. The layout is shown in Figure 12. It provides links to numerical weather prediction output from different centers, including the UK Met Office and ECMWF. It also proves an outlook for significant weather events for southern Africa which is compiled by forecasters in South Africa for five days ahead.

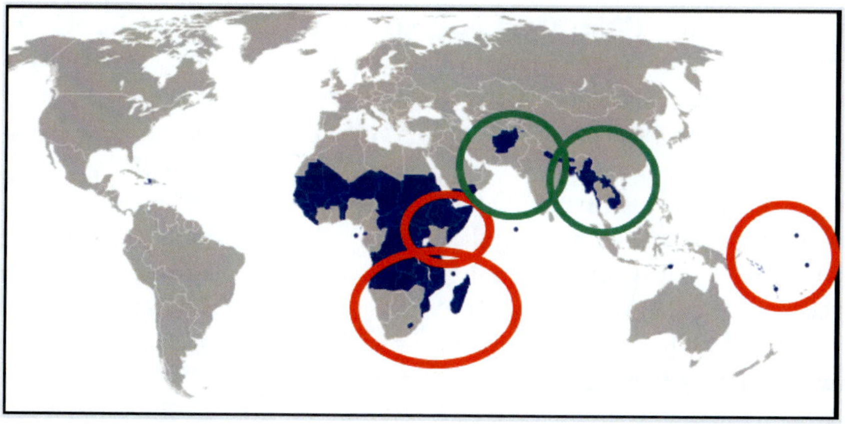

Figure 11. Existing SWFDP (in red) and planned SWFDP (in green) overlaid on the LDC (as in Figure 1).

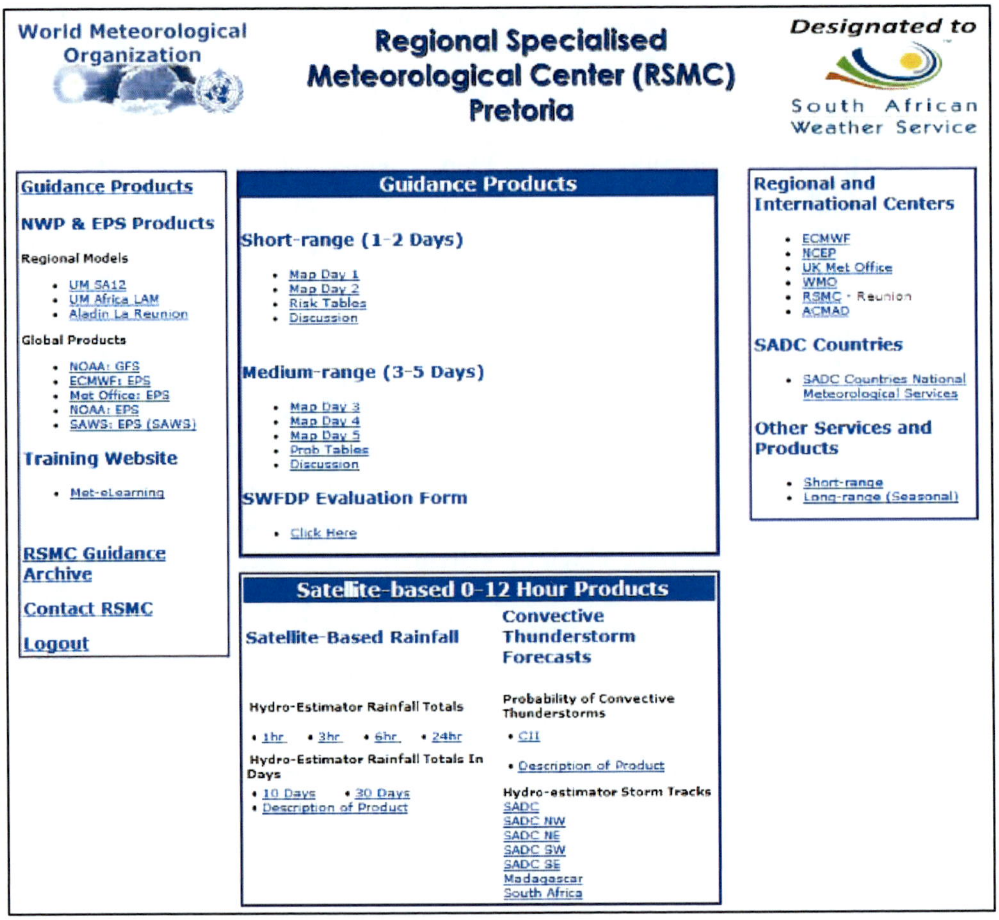

Figure 12. Layout of the South African SWFDP information as portrayed on the RSMC web portal.

Since the implementation of the Hydroestimator at SAWS, accumulation products of the Hydroestimator for 1 hour, 3 hours, 6 hours, 24 hours, 10 days and 1 month have been added to the SWFDP website where the NMHS of southern African countries have access to it.

The real time tracking of the 10 mm per hour rainfall cores is also available for different smaller regions over the sub-continent. Three-hourly-time averages of the Combined Instability Index, which are updated every hour, are available on the southern African SWFDP website to provide information on the probability of convective activity over this region.

Currently, the eastern African SWFDP is in the development phase and the southern African SWFDP is assisting to get a similar website as well as similar and relevant products set up for this area. Products specific to the eastern African region will be developed in due course especially for Lake Victoria, which is subject to severe convective events leading to loss of life. The WMO working groups on nowcasting, mesoscale modeling, as well as SWFDP development, will collaborate in this effort.

Conclusion

Forecasting for the first 12 hours ahead can be a challenging task for most forecasters. If this task has to be performed without the luxury of powerful tools, such as ground based weather radars systems, as is the case in most DC and LDC, it is an even more daunting task. In the absence of expensive and extensive observation networks and remote sensing tools, nowcasting and very short range forecasting systems for developing countries and/or least developing countries should focus on numerical weather prediction and satellite data, as well as other observation data, to address their needs.

The use of satellite products and satellite and model blended products is one way of addressing the needs in this forecasting time frame since MSG data and imagery are available every 15 minutes. Using the specialized products and applications supplied by the EUMETSAT MPEF and the various SAF can open a wide range of possibilities. Many recent tools for the pre-convective, convection initiation and mature stages of convection have been developed since 2005. Although most of these algorithms can be useful and applicable in DC and LDC, some of the methods require extensive programming and computer resources which are usually limited in DC and LDC. Despite the fact that some products (MPEF) are readily disseminated to all countries which receive the MSG data via the EUMETcast system, it is still beyond the capability of most DC and LDC to decode and visualize these products. Some applications have been developed in South Africa, and these are disseminated to southern African countries via the internet in graphical format.

It was shown that satellite-based instability indices can provide information of excellent temporal and spatial resolution to learn more about and improve forecasts of the possibility of convection with a few hours lead time. The use of products such as the Global Instability Indices or Combined Instability Index proved to be worthwhile for countries without radar systems to improve nowcasting and very short-range forecasting of convection. Satellite-based precipitation estimation can also be useful to indicate, in real-time, where it is raining and approximately how much, even in areas not covered by other sensors. Accumulation and tracking of satellite-based precipitation estimation products can aid in assessing where flooding and/or flash flooding can occur due to slow moving storms. For countries where both convection and stratiform precipitation occur, a product combining satellite-based convective rainfall estimation with stratiform rainfall from numerical weather prediction models can be of use, for example, along the coastlines of South Africa.

Satellite-based precipitation estimations provide valuable input to the WMO Flash Flood Guidance System (FFGS) projects in various developing regions of the world. The Southern African Regional Flash Flood Guidance (SARFFG) project is highly dependent on satellite-based precipitation estimation to determine which small catchments have the potential to experience flash floods. This system is web based and thus provides easy access to all DC and LDC on the subcontinent.

The Severe Weather Forecasting Demonstration Projects (SWFDP) can serve as an excellent and easy way to disseminate information via the internet instead of requiring expensive visualization software. NMHS in southern Africa are convinced that the SWFDP information has improved warning services. With the move to satellite-based products for future SWFDP development, it is envisaged that more emphasis will be on satellite data to complement the inputs from numerical weather prediction models. A recommendation can be that a product for severe convection gets added to the suite of products to assist forecasters in the prevention of damage to property and loss of life through this method of dissemination.

REFERENCES

Berendes, T.A.; J.R. Mecikalski; W.M. MacKenzie Jr; K.M. Bedka; U.S. Nair. Convective cloud identification and classification in daytime satellite imagery using standard deviation limited adaptive clustering. *J. Geophys. Res.*, 2008, 113, D20207, doi: 1029/2008JD010287.

Charba, J. P. Operational System for predicting thunderstorms two to six hours in advance. Report: NOAA Technical Memorandum NWS TDL-64. National Oceanic and Atmospheric Administration, US Department of Commerce, 1977, 24pp.

De Coning, E.; J. Kerkmann; M. Setvak. (2009) Severe convective storms over the southeastern parts of South Africa (27 November 2009). EUMETSAT Case study web page: http://oiswww.eumetsat.org/WEBOPS/iotm/iotm/20091127_convection/2009 1127 _convection.html.

De Coning, E.; E. R. Poolman. South African Weather Service operational satellite based precipitation estimation technique: applications and improvements. *Hydrol. Earth Syst. Sci.*, 2011, 15, 1131-1145, Available at: www.hydrol-earth-syst-sci.net/15/1131/2011/ doi:10.5194/hess-15-1131-2011.

De Coning, E.; M. Koenig; J. Olivier. The Combined Instability Index: A new very-short range convection forecasting technique for southern Africa. Meteorological Applications, 2011, 18, 421-439. Available: http:// onlinelibrary.wiley.com/doi/10.1002/met. 234/abstract.

Dixon, M. J.; G. Wiener. Thunderstorm Identification Tracking Analysis and Nowcasting - A radar based methodology. *J. Atmos. Ocean. Tech.*, 1993, 10(6), 785-797.

Dostalek, J. F.; T.J. Schmit. Total precipitable water measurements from GOES sounder derived product imagery. *Wea. Forecasting*, 2001, 16, 573-587.

Forster, C.; E. de Coning; S. Diebel; T. Zinner. Satellite-based thunderstorm tracking, monitoring and nowcasting over South Africa. EUMETSAT Meteorological Satellite Conference, 20-24 September 2010, Córdoba, Spain EUMETSAT P.57, ISBN 978-92-9110-089-7, ISSN 1011-3932.

Georgakakos, K. P. Mitigating Adverse Hydrological Impacts of Storm on a Global Scale with High Resolution: Global Flash Flood Guidance. Proceedings of the International Conference on Storms, Storms Science to Disaster Mitigation, Brisbane, Australia, 5-9 July 2004.

Georgakakos, K. P. Analytical Results for Operational Flash Flood Guidance. *Journal of Hydrology*, 2006, 317, 81-103.

Georgiev, C. G. Use of Data from the Meteosat water vapor channel and surface observations for studying pre-convective environment of a tornado producing storm. *Atmos. Res.*, 2003, 67-68, 231-246.

Global Humanitarian Forum. (2008). Weather information for all initiative 2008-2012. Available at: http://info.publicintelligence.net/WIFA_Project_Outline_Executive_Summary.pdf.

Hayden, C. M. GOES-VAS simultaneous temperature-moisture retrieval algorithm. *J. Appl. Meteorol.*, 1988, 27, 705-733.

Koenig, M. Atmospheric Instability parameters derived from MSG SEVIRI observations. (Report: Technical Memorandum No.9). 2002, EUMETSAT.

Koenig, M. The Global Instability Indices Product Algorithm Theoretical Background Documents. (Report: EUM/MET/REP/07/0164 v1), 2007, EUMETSAT.

Koenig, M.; E. de Coning. MSG for nowcasting – Experiences over Southern Africa. EUMETSAT Meteorological Satellite Conference, Helsinki, Finland. 12-16 June 2006. EUM P.48, ISBN 92-9110-076-5.

Koenig M.; E. de Coning. The MSG Global Instability Indices Product and its use as a nowcasting tool. *Wea. Forecasting*, 2009, 24, 272-285.

Kondragunta, C. Quantitative Precipitation Estimation in the National Weather Service. NOAA's Water Resources Information. Office of Systems Development. NOAA/NESDIS. George Washington University, DC. March 13 2007. George Washington University website: http://www.gwu.edu/~spi/assets/docs /Chandra_Kondragunta-NOAA's%20Water%20Resources%20Information.pdf.

Kruger, A. C. Climate of South Africa, Precipitation. (Report No. WS47). Available at South African Weather Service, 2007, 41pp.

Mecikalski, J.R.; K. M. Bedka. Forecasting Convective initiation by monitoring the evolution of moving convective daytime GOES imagery. *Mon. Wea. Rev.*, 2006, 134, 49-78.

Mecikalski, J. R.; J. J. Murray; W. F. Feltz; D. B. Johnson; K. M. Bedka; S. T. Bedka; A. J. Wimmers; M. Pavolonis; T. A. Berendes; J. Haggerty; P. Minnis; B. Bernstein; E. Williams. Aviation applications for satellite-based observations of cloud properties, convective initiation, in-flight icing, turbulence and volcanic ash. *Bull. Amer. Meteor. Soc.*, 2007, 88, 1589-1607.

Mecikalski, J. R.; W. M. Mackenzie; M. Koenig; S. Muller. Use of Meteosat Second Generation Infrared Data in 0-1 hour Convective Initiation Nowcasting. Part 1.Infrared Fields. *J. Appl. Meteor. Climate*, 2010, 49, 521-534.

Morgan, J. Applications of Meteosat Second Generation. Report No. EUM BR 11, 2002, EUMETSAT.

Peppler, R. A. A review of static stability indices and related thermodynamic parameters. Report: SWS Misc. Publ. 104, Illinois State Water Survey Division, Climate and Meteorology Section, 1988, 94 pp.

Petersen, R. A.; R. Aune; T. Rink. Objective short range forecasts of the pre-convective environment using SEVIRI data. EUMETSAT Meteorological Satellite Conference, 20-24 September 2010, Córdoba, Spain EUMETSAT P.57, ISBN 978-92-9110-089-7, ISSN 1011-3932.

Poolman, E. R.; H. Chikoore; F. Lucio. (2008). Public benefits of the Severe Weather Forecasting Demonstration Project in south-eastern Africa. WMO Newsletter MeteoWorld, Available at: http://www.wmo.int/pages/publications/meteoworld/archive /dec08/index_en.html.

Poolman, E. R. The development of the South African Flash Flood Guidance System (SAFFG). SAWS Internal Report, 2012. Available at South African Weather Service, Pretoria, South Africa.

Rabin R.; S. F. Corfidi; J. C. Brunner; C. E. Hain. Detecting winds aloft from water vapor satellite imagery in the vicinity of storms. *Weather*, 2004, 59, 149-166.

Rosenfeld, D.; W. L. Woodley; A. Lerner; G. Kelman; D. T. Lindsey. Satellite Detection of severe convective storms by their retrieved vertical profiles of Cloud Particle Effective Radius and Thermodynamic Phase. *J. Geophys. Res.*, 2008, 113, D04208, doi:10.1029/2007JD008600.

Schmetz, J.; R. Borde; K. Holmund; M. Koenig. Upper tropospheric divergence in tropical convective systems from Meteosat-8. *Geophys. Res. Lett.*, 2005, 32, L24804, doi: 10.1029/2005GL024371.

Schmit, T. J.; W. F. Feltz; W. P. Menzel; J. Jung; A. P. Noel; J. N. Heil; J. P. Nelson; G. S. Wade. Validation and use of GOES sounder moisture information. *Wea. Forecasting*, 2002, 17, 139-154.

Scofield, R. A.; R. J. Kuligowski. Status and outlook of operational satellite precipitation algorithms for extreme-precipitation events. *Wea. Forecasting*, 2003, 18, 1037-1051.

Setvak, M.; D. T. Lindsey; P. Novak; P. K. Wang; M. Radova; J. Kerkmann; L. Grasso; S. Su; R. M. Rabin; J. Stastka; Z. Charvat. Satellite-observed Cold-ring-shaped Features atop Deep Convective Clouds. *Atmos. Research*, 2010, 97, 80-96, doi:10.1016/j.atmosres.2010.03.009.

STAR Satellite Rainfall Estimates. Center for Satellite Applications and Research website: http://www.star.nesdis.noaa.gov/smcd/emb/ff/index.php.

UN Secretary-General Report 2012 on the Implementation of the Istanbul Program of Action for Least Developed Countries for the Decade 2011-2020. Available at: http://www.unohrlls.org/UserFiles/File/UN_LDC_Factsheet_053112.pdf.

Vicente, G.; R. A. Scofield; W. P. Mentzel. The operational GOES infrared rainfall estimation technique. *Bull Amer. Meteor. Soc.*, 1998, 79, 1883-1898.

Wagner, T. J.; W. F. Feltz; S. A. Ackerman. The temporal evolution of convective indices in storm producing environments. *Wea. Forecasting*, 2008, 23, 786-794.

Wilks D. S. Statistical Methods in Atmospheric Sciences, Second Edition. Elsevier Science and Technology Books. 2005, ISBN-13: 9780127519661. 627pp.

World Meteorological Organization. (2012) Commission for Basic Systems. CBS Steering Group Severe Weather Forecasting Demonstration Project. Fourth Meeting. Geneva, Switzerland, 28 February – 2 March. http://www.wmo.int/pages/prog/www/CBS-Reports/documents/SG-SWFDP-4_Final_Report.pdf.

Zinner, T.; H. Mannstein; A. Tafferner. Cb-TRAM: Tracking and Monitoring Severe Convection from onset over Rapid Development to Mature Phase Using Multichannel Meteosat-8 SEVIRI Data. *Meteor. and Atmos. Phys.*, 2008, 101, 191-210.

In: Recent Advances in Satellite Research and Development ISBN: 978-1-62417-443-8
Editors: Samuel Gardiner and Kenneth P. Olsen

Chapter 4

POSSIBILITIES OF ELIMINATING IONOSPHERIC EFFECTS IN MULTIFREQUENCY GLOBAL NAVIGATION SATELLITE SYSTEMS

***M. V. Tinin*[1,*], *B. C. Kim*[2] *and E. V. Konetskaya*[1]**
[1]Irkutsk State University, Irkutsk, Russia
[2]Pusan National University, Samrangjin-eup,
Miryang-city, Gyeoungnam, Korea

ABSTRACT

Methods for environmental diagnostics with the use of global navigation satellite systems (GNSS) have been progressing intensively. Failures of these methods are largely caused by the anisotropic, space-time-variable ionospheric plasma that is in the path of a satellite signal. The ionospheric effects are usually removed using dispersive features of the ionospheric plasma. To do it, phase or/and time delays of a satellite signal is measured at two frequencies. The dual-frequency measurements neutralize the major part of the ionospheric error (the first-order error). But this accuracy is already insufficient for high-precision measurements needed for monitoring the lower atmosphere to forecast the weather and climatic changes and for examining the deformation of the Earth's surface to predict earthquakes and volcanic activity. This chapter summarizes findings of the investigations into the residual ionospheric error. Here we mainly rely on authors' publication supplemented by literature data and some unpublished findings. In the geometrical optics approximation, the errors of phase and group paths of an ionospheric radio wave at GNSS frequencies may be represented as series in inverse powers of frequency. It is possible to avoid the commonly encountered inaccuracies here, if wave propagation of certain (right) polarization is considered instead of ordinary or extraordinary waves as in traditional investigations. We show how the dual-frequency measurements can be exploited to remove not only the first-order error, but also the second-order error associated with the geomagnetic effect on the refractive index in the ionospheric plasma. After eliminating the second-order error in dual-frequency measurements, we can remove third-order effects associated with ray bending in the

* E-mail: mtinin@api.isu.ru.

inhomogeneous ionosphere, using triple-frequency measurements. It is shown that in addition to the background ionosphere a considerable (often dominating) contribution to this error is made by random ionospheric inhomogeneities. Effects of ionospheric inhomogeneities in triple-frequency GNSS can be eliminated only for inhomogeneities with transversal scales exceeding the Fresnel radius. Diffraction makes it impossible to remove the effects of inhomogeneities with smaller scales. This leads to the residual error in triple-frequency measurements. We discuss possibilities of existing and currently developed spatial signal processing methods being able to eliminate diffraction effects and improve accuracy of multifrequency GNSS measurements.

1. Introduction

Almost simultaneously with first satellite launchings, the US and USSR military departments started working on exploiting these satellites for navigation [Hooijberg, 2008; Bonnor, 2012]. First navigation satellite systems were designed as far back as the early 1970s. In the 1990s, further development in the systems brought such global navigation satellite systems (GNSS) as GPS in the USA and GLONASS in Russia into being. Moreover, the European system GALILEO, the Chinese system COMPASS, and a number of other systems are being developed nowadays [Harper, 2010]. These systems quickly came into general use not only for military commitments, but also in many fields of economy and science [French, 1996; El-Rabbany, 2002]. The reason is that the space reference marks (Figure 1) are easy to apply for fixing positions almost all over the Earth's surface and ensure high-precision measurements. Coordinates of a point are determined from results of measurements at this point of phase or envelope (code) delay of signals emitted at least by four satellites. Obviously that the positional accuracy largely depends on the accuracy of estimating the distance to these satellites by measuring phase and code paths (pseudo-range) of a GNSS signal.

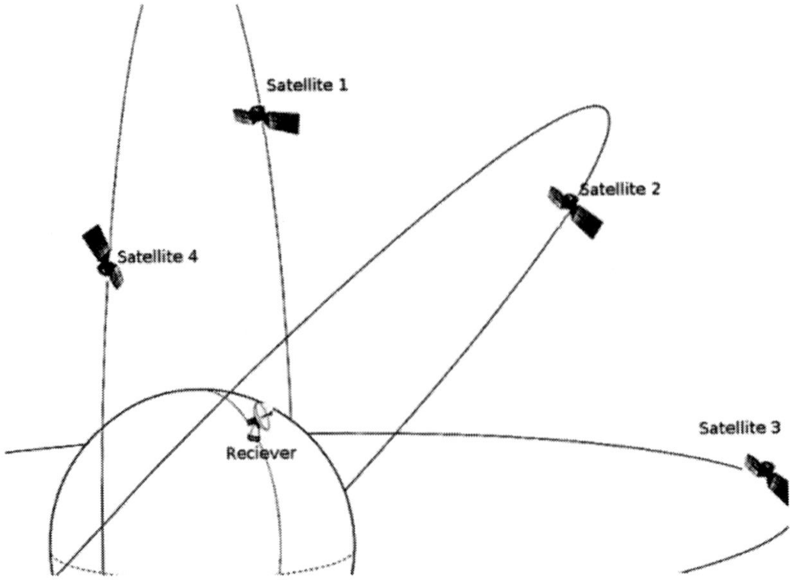

Figure 1. The schematic drawing of relative positions of GNSS satellites and a ground-based receiver.

Observation equations for the measured pseudodistance P and phase path φ take the form [Langley, 1998a]:

$$P = D - (\delta t_r - \delta t_s)c + P_{ion} + \delta_{tro} + \varepsilon_p , \tag{1}$$

$$\varphi = D - (\delta t_r - \delta t_s)c + \lambda N_\lambda + \varphi_{ion} + \delta_{tro} + \varepsilon_\Phi , \tag{2}$$

where D is the geometric distance between the satellite and the receiver; c is the velocity of light; δt_r and δt_s denote the clock errors of the receiver and satellite, respectively; P_{ion} and φ_{ion} are the ionospheric errors of code and phase measurements; δ_{tro} is the tropospheric error; N_λ is the integer resulting from the phase measurement ambiguity; ε_P and ε_Φ include such residual effects as multipath propagation, noises in reception and transmission paths, etc. Note that Langley [1998a] suggests that $P_{ion} = -\varphi_{ion}$, but you will see below that this is not exactly correct.

For definiteness, the additional terms to the geometric distance on the right in Eq. (1)–(2) will be referred to as measurement errors, and vice versa their values reversed in sign needed to be added to the measurement results to obtain the true distance will be termed corrections.

As for now, the instrumentation errors have largely been eliminated or are corrected to a high accuracy; the potential accuracy of the most accurate phase measurements may be as high as 1–2 mm [Alber et al, 1997; Langley, 1998a]. Therefore at present many researchers pay considerable attention to the terms P_{ion}, φ_{ion}, δ_{tro}, accounting for the change of phase and code delays caused by the Earth's atmosphere. The largest contribution (up to 100 m) is made by the upper atmosphere called the ionosphere [Kunches and Klobuchar 2001; Grewal et al, 2007; Datta-Barua et al, 2008]. When determining the role the ionosphere plays in the GNSS measurements, account must be taken of its strong spatial-temporal variability and ionospheric plasma dispersion revealing itself in the radiation frequency dependence of its permittivity. The first peculiarity strongly complicates the consideration of the ionospheric effects via simulation. Thus the most popular algorithm for ionospheric correction based on the Klobuchar model (see [Grewal et al, 2007; Tsui, 2005; Seeber, 2003; Xu, 2007]) eliminates only 50% of the ionospheric error [Langley, 1998b; Tsui, 2005]. There are other models for correcting the ionospheric error (see for example, [Ivanov et al, 2011]), but they do not considerably diminish this error as well.

Another, so-called differential, approach [Kennedy, 2002; Seeber, 2003; Grewal et al, 2007; Samama, 2008], is founded on the interpolation of ionospheric corrections determined at reference stations. The accuracy of such methods depends on the correlation radius of corrections, the interpolation method, and accuracy of the reference station measurements which can mostly be obtained through dual-frequency observations.

Off-line (independent of any reference stations) and high-precision dual-frequency GNSS measurements (see [Grewal et al, 2007; Tsui, 2005; Seeber, 2003; Xu, 2007]) exploit dispersion features of ionospheric plasma to eliminate its effects when measuring signal phase or delay at two frequencies. The dual-frequency measurements can remove up to 99% of the ionospheric error [Klobuchar and Kunches, 2003] and thus improve the positioning accuracy to dm level. This accuracy is sufficient for many applications. However, the above mentioned potential accuracy of GNSS measurements equal to 1–2 mm may be necessary to solve new problems.

First of all, it is associated with new possibilities of the troposphere monitoring with the goal of improving the accuracy of weather and climatic change forecasts [Kursinski et al., 1997]. Excluding the ionospheric effect, from (1) or (2) we can isolate the tropospheric component δ_{tro}, that is closely related to the lower atmosphere temperature and density. Of particular interest for such explorations are occultation observations when a signal is measured aboard Low Earth Orbit (LEO) satellite rather than at ground-based stations. As the LEO satellite orbit is lower than that of the GNSS satellite, its speed is much higher, and we can track its radio setting. Under certain conditions, this scanning gives both ionospheric and tropospheric profiles. Also important is that the measurements provide rapid information from places hardly accessible for common meteorological research methods. This problem requires improved accuracy because the accuracy of order of dm may be of order of the effect considered.

The second problem that needs more precise GNSS measurements is the observation of the motion of lithospheric plates and deformations of the land surface forerunning and accompanying volcanic activity and earthquakes [Schaal and Larocca, 2001].

These problems stimulate researchers' interest in analyzing dual-frequency measurement errors and searching for methods of their elimination in multifrequency GNSS.

This Chapter presents an analysis of the current status of investigations conducted to improve measurement accuracy in multifrequency GNSS. Here we will mainly follow authors' publications supplemented by literature data and some unpublished findings. Besides, here we analyze only the ionospheric error in (1)–(2), and therefore will omit the subscript, assuming $\varphi = \varphi_{ion}$ and $P = P_{ion}$.

The subsequent contents include the following. In Section 2, the ionospheric error and ways of eliminating it are analyzed with the aid of geometrical optics (GO). Subsection 2.1 presents an expression for the phase path in the GO approximation, which is valid for the right-handed polarized wave of a GNSS signal independently of whether this wave is ordinary or extraordinary. Subsection 2.2 represents the phase path as series in inverse powers of frequency. The series' terms depending on frequency f as f^{i+1} are i-order corrections. Some characteristics of these corrections are discussed. Subsection 2.3 describes ways of eliminating errors of different orders in dual- and triple-frequency GNSS. Section 3 is concerned with the consideration of diffraction effects in GNSS measurements. In Subsection 3.1, the second-order Rytov approximation has been used to obtain ionospheric corrections of different orders with due regard to diffraction effects. In Subsection 3.2, the numerical simulation shows that ionospheric irregularities of scales less than the Fresnel radius in the spectrum makes it impossible to remove the first- and third-order ionospheric errors completely. Section 4 summarizes the findings of the investigation and discusses possible ways of eliminating diffraction effects to improve accuracy of multifrequency GNSS measurements.

2. Ionospheric Errors in GNSS Measurements at the Geometrical Optics Approximation

In this Chapter, we will principally deal with phase measurements; firstly, because they have a higher accuracy than propagation time measurements; secondly, the phase path dependence $\varphi(f)$ on the frequency f makes easier the determination of a group path P from the relation

$$P = \frac{\partial}{\partial f}\left(f\varphi(f)\right). \quad (3)$$

2.1. The Geometrical Optics Approximation for a Signal Phase Path in the Anisotropic Inhomogeneous Ionospheric Plasma

Let us examine propagation of a monochromatic wave with the time dependence $e^{-i\omega t}$ in the inhomogeneous ionosphere. As is well-known [Kravtsov and Orlov, 1990], in the GO approximation from eikonal equation

$$\left(\nabla\varphi_g\right)^2 = \varepsilon\left(\mathbf{r}, f\right) = n^2\left(\mathbf{r}, f\right) \quad (4)$$

one can easily find the phase path (eikonal)

$$\varphi_g\left(f\right) = \int_0^{s_t} n\left(\mathbf{r}\left(s, f\right), f\right) ds, \quad (5)$$

where the integration is done along the ray path $\mathbf{r}\left(s, f\right)$, satisfying equation system

$$\frac{d\mathbf{r}}{ds} = \frac{\mathbf{p}}{n\left(\mathbf{r}, f\right)}, \qquad \frac{d\mathbf{p}}{ds} = \nabla n\left(\mathbf{r}, f\right), \quad (6)$$

$$\mathbf{r}\left(0, f\right) = \mathbf{r}_0, \qquad \mathbf{r}\left(s_t, f\right) = \mathbf{r}_t. \quad (7)$$

Here $\mathbf{r}_0$ и $\mathbf{r}_t$ are coordinates of the source and the receiver, $n\left(\mathbf{r}, f\right)$ is the refractive index which for a collisionless ionospheric plasma is determined by the following formula [Ginzburg, 1970; Budden, 1985]:

$$n^2 = 1 - \frac{X}{1 + i\rho Y \cos\theta} \quad (8)$$

In Eq. (8), we use common notations: $X=\frac{\omega_p^2}{\omega^2}=\frac{f_p^2}{f^2}=\frac{80.6N}{f^2}$, $Y=\frac{\omega_H}{\omega}=\frac{f_H}{f}=\left|\frac{eB_0}{2\pi mf}\right|$, $f_p=\sqrt{80.6N}$ is the plasma frequency, N is the electron density, $f_H=\left|\frac{eB_0}{2\pi m}\right|$ is the electron gyrofrequency, e and m are the electron charge and mass at rest, $\mathbf{B}_0(\mathbf{r})$ is the geomagnetic field, θ is the angle between $\mathbf{B}_0$ and wave propagation direction; the polarization ρ is determined by relation [Ginzburg, 1970; Budden, 1985]:

$$\rho=\frac{0.5iY\sin^2\theta\pm i\sqrt{0.25Y^2\sin^4\theta+\cos^2\theta(1-X)^2}}{(1-X)\cos\theta}. \tag{9}$$

When compared (8)-(9) to [Ginzburg, 1970; Budden, 1985] it is necessary to consider that here the time dependence is $e^{-i\omega t}$ as opposed to $e^{i\omega t}$ and the determination of the angle θ differs by π from that accepted in [Ginzburg, 1970; Budden, 1985].

Substituting (9) in (8) gives the well-known Appleton-Hartree formula [Ginzburg, 1970; Budden, 1985]:

$$n^2=1-X\left[1-\frac{Y^2\sin^2\theta}{2(1-X)}\mp\sqrt{\frac{Y^4\sin^4\theta}{4(1-X)^2}+Y^2\cos^2\theta}\right]^{-1}. \tag{10}$$

If the branch of the root in (9)–(10) is held fixed by condition

$$\mathrm{Re}\sqrt{\frac{Y^4\sin^4\theta}{4(1-X)^2}+Y^2\cos^2\theta}>0, \tag{11}$$

the lower (+) sign before the root denotes the ordinary wave; the upper (-) sign is related to the extraordinary wave. This definition is associated with the emergence of resonance effects in the extraordinary wave at the frequencies close to the gyrofrequency, i.e. at $f\to f_H$. GNSS signal frequencies are far from the gyrofrequency, and the difference between ordinary and extraordinary waves is only in a slight excess of ionospheric effects for the extraordinary wave.

A GNSS signal generally has the right-hand polarization. A number of papers [Brunner and Gu, 1991; Bassiri and Hajj, 1992, 1993; Hoque and Jakowski, 2006, 2008a] reason that this polarization belongs only to the extraordinary wave. Moore and Morton [2011] prove this suggestion to be wrong. Both extraordinary and ordinary waves do have the right-hand polarization, depending on wave propagation direction relative to the magnetic field [Budden, 1985]. Thus, to avoid confusion it is more convenient to associate the GNSS signal with right-hand polarized wave rather than with extraordinary or ordinary waves. To do this, we use the fact that in the GNSS frequency range (1.2–1.5 GHz) in (8)–(10) there is a small parameter ν inversely proportional to the square of the radiation frequency:

$$\nu = \left(f_p / f\right)^2 = O\left(10^{-4}\right) << 1. \qquad (12)$$

Given this, for the right-hand polarization we have from (9)

$$\rho = i \qquad (13)$$

and obtain from (10)

$$n^2 = 1 + \varepsilon_1 , \qquad (14)$$

where

$$\varepsilon_1 = -X\left[1 + Y\cos\theta\right] + O\left(X^2\right) + O\left(XY^2\right). \qquad (15)$$

This quasi-longitudinal approximation is fairly accurate for GNSS frequencies [Brunner and Gu, 1991].

2.2. Ionospheric Corrections of Different Orders

Ionospheric effects in transionospheric propagation were analysed by [Brunner and Gu, 1991; Ashmanets et al., 1996; Strangeways and Ioannides, 2002] using numerical integration of equations (6)–(7).

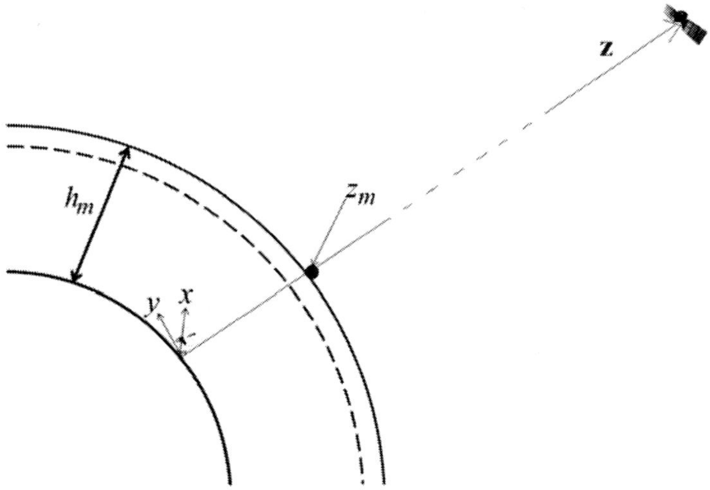

Figure 2. The coordinate system employed in calculations of ionospheric corrections.

Note that such simulation is a very complicated, two-point problem because it is necessary to solve Eqs. (6) – (7) and to use a homing-method adjusting the solution to the observation point with millimeter accuracy at a distance of over 20000 km. While in frequencies 20–200 MHz [Ashmanets et al., 1996] such efforts are justified, in the L-band, exploited in state-of-the-art GNSS, methods based on numerical simulation of strict ray equations (6)–(7) have not received wide acceptance yet. In this band, the phase path can be found from equations (5)–(7) far easier with the aid of the perturbation theory.

Let us choose a coordinate system $\mathbf{r}=\{x,y,z\}=\{\boldsymbol{\rho},z\}$ with the z-axis passing through radiation $\mathbf{r}_0=\{0,0,z_0\}$ and reception $\mathbf{r}_t=\{0,0,z_t\}$ points (see Figure2). Since owing to (12) $\varepsilon_1=O(\nu)$ is a small quantity, the ray path slightly deviates from the straight line $\boldsymbol{\rho}(z)=0$ and this deviation $\boldsymbol{\rho}(z)$ may be found by solving ray equations (6)–(7) with the aid the perturbation theory [Tinin et al., 2004; Kim et al., 2005; Kim and Tinin, 2006; Gherm et al., 2006; Kim and Tinin 2007a]. Substituting this solution in (5) and taking into account that expansion

$$n=\sqrt{\varepsilon}=\sqrt{1+\varepsilon_1}\approx 1+\varepsilon_1/2-\varepsilon_1^2/8+... \tag{16}$$

is valid when (12) is fulfilled we can obtain [Kim and Tinin 2007a]:

$$\varphi_g=D+\frac{1}{2}\int_{z_0}^{z_t}\varepsilon_1(0,0,z')dz'-\frac{1}{8}\int_{z_0}^{z_t}\varepsilon_1^2(0,0,z')dz'-\frac{1}{2}\int_{z_0}^{z_t}p_1^2(0,0,z')dz'+O(\nu^3), \tag{17}$$

$$\mathbf{p}_1=\frac{1}{2}\int_{z_0}^{z_t}\frac{\partial}{\partial\boldsymbol{\rho}}\varepsilon_1(0,0,z')dz'. \tag{18}$$

From now on we will consider that, as the satellite is at a much more distance from the ionosphere than the observer, the incident wave can be taken as the plane wave. Substituting (15) in (17)–(18) yields [Kim and Tinin 2007a, 2011]:

$$\varphi_g(f)=D_g^{(1)}\approx D-\Delta D_{1g}^{(1)}(f)-\Delta D_{2g}^{(1)}(f)-\Delta D_{3g}^{(1)}(f), \tag{19}$$

where i-order single-frequency corrections $\Delta D_{ig}^{(1)}(f)$ in the GO approximation are given by the following expressions:

$$\Delta D_{1g}^{(1)}(f)=40.3f^{-2}I_{1g}\;, \tag{20}$$

$$\Delta D_{2g}^{(1)}(f)=40.3f^{-3}I_{2g}\;, \tag{21}$$

$$\Delta D_{3g}^{(1)}(f) = (40.3)^2 f^{-4}\left[I_{3bn} + I_{3sq}\right]. \tag{22}$$

here

$$I_{1g} = \int_{z_0}^{z_t} N(z')dz' \tag{23}$$

is the total electron content (TEC),

$$I_{2g} = \int_{z_0}^{z_t} N(z')f_H(z')\cos\theta(z')dz', \tag{24}$$

$$I_{3bn} = \frac{1}{2}\int_{z_0}^{z_t}\left[\int_{z_0}^{z}\frac{\partial N(\boldsymbol{\rho}, z')}{\partial \boldsymbol{\rho}}\bigg|_{\boldsymbol{\rho}=0} dz'\right]^2 dz, \tag{25}$$

$$I_{3sq} = \int_{z_0}^{z_t} N^2(z')dz'. \tag{26}$$

Another way of finding ionospheric corrections involves the direct solution of eikonal equation (4) with the use of the perturbation theory without solving ray equations [e.g. Moritz, 1967; de Munck, 1970; Williams, 1975; Gu and Brunner, 1990]. As is shown by Kim and Tinin [2007a], this approach also results in (19)–(26).

The main contribution to the ionospheric effects is made by the TEC-proportional first-order correction (20). This correction is very important for single-frequency observations; therefore it has been the objective of much research.

In examining effects of ionospheric plasma in radio wave propagation, it is convenient to represent its electron density N as sum of two components

$$N(\mathbf{r}) = \bar{N}(\mathbf{r}) + \tilde{N}(\mathbf{r}), \tag{27}$$

where $\bar{N}(\mathbf{r})$ is the average electron density (density of the large-scale background ionosphere); $\tilde{N}(\mathbf{r})$ is the random component of this density. The spatial variability of the background ionosphere $\bar{N}(\mathbf{r})$ is associated with its quasi-layered structure, and the random component $\tilde{N}(\mathbf{r})$ accounts for the ionospheric plasma turbulence. After substituting (27) in (20), (23) correction $\Delta D_{1g}^{(1)}$ is also divided into two components: mean $\Delta\bar{D}_{1g}^{(1)}$ and random $\Delta\tilde{D}_{1g}^{(1)}$. The first-order correction fluctuations $\Delta\tilde{D}_{1g}^{(1)}$ are defined by thoroughly studied phase fluctuations (phase scintillations). Observation of the phase fluctuations is used in studies of ionospheric irregularities [e.g. Gershman et al., 1984]. Besides, it is well known that they affect the GNSS operational stability [e.g. Afraimovich et al. 2011].

Considering the small value of the second-order effects and that usually $\bar{N}(\mathbf{r}) >> \tilde{N}(\mathbf{r})$, we may ignore their fluctuations when calculating these effects and assume

$$\Delta D_{2g}^{(1)}(f) \approx \Delta \bar{D}_{2g}^{(1)}(f) = 40.3 f^{-3} \bar{I}_{2g} \quad = 40.3 f^{-3} \int_{z_0}^{z_t} \bar{N}(z') f_H(z') \cos\theta(z') dz' \tag{28}$$

Second-order correction (28) is associated with the geomagnetic field effect on refractive index (10). Currently it has been the objective of much research. Bassiri and Hajj [1992, 1993] noticed that, since the scale of the geomagnetic field $\mathbf{B}_0(\mathbf{r})$ exceeds that of the electron density $\bar{N}(\mathbf{r})$, one can take values of function $f_H(z')\cos\theta(z')$ out of the integral in (28) at a point $z' = z_m$ on the ray path connecting a GNSS satellite with a ground-based user. A point of intersection of the ray with the ionospheric layer maximum (see Figure2) is normally taken for z_m. Thus, (28) yields the following approximation

$$\Delta D_{2g}^{(1)}(f) \approx 40.3 f_H(z_m) \cos\theta(z_m) f^{-3} \int_{z_0}^{z_t} \bar{N}(z') dz' = 40.3 f_H(z_m) \cos\theta(z_m) f^{-3} I_{1g} . \tag{29}$$

So, to determine the second-order correction, one need know only TEC I_{1g}, magnetic field model, satellite and observer coordinates. Moreover, because the second-order effects are weak, there is no need for high requirements on the accuracy of assignment of these coordinates.

The accuracy of thin layer approximation (29), as evident from the numerical simulation results [Bassiri and Hajj, 1992, 1993], is quite high.

In Figure 3 is an example of the results of calculations (see also [Tinin and Konetskaya, 2011]) of the error in estimating the second-order correction in thin layer approximation (29) for the satellite elevation angle of 30° and the azimuth of 180°. In the calculations, we exploited GPS frequency $f_1 = 1{,}575.42 \times 10^6\, Hz$ and a dipole geomagnetic field model with the strength of 0.312 μT at the geomagnetic equator. For the ionospheric model we chose Chapman layer

$$\bar{N}(r) = \frac{f_c^2}{80.6} \exp\left\{0.5\left[1 - \zeta(r) - \exp(\zeta(r))\right]\right\}, \tag{30}$$

where $\zeta = (r - h_m - R_e)/H$, r is the distance to the Earth center, $R_e = 6370 km$ is the Earth's radius, $h_m = 350 km$ is the height of the ionospheric layer maximum,

$H = 70km$ is the typical scale of the layer, $f_c = 14\times10^6\,Hz$ is the critical frequency of the layer. These parameters correspond to the total electron content with the vertical propagation (VTEC) equal to 70 TECU ($1\ \text{TECU} = 10^{16}\,\text{electrons}/\text{m}^2$). As observed in the Figure3, in this case the error is within 0.5 mm. As the satellite elevation angle changes from 10° to 90°, the error in calculating the second-order correction decreases in modulus. At the vertical position of the satellite over a receiver, the error is within 0.2 mm. The simulation at an elevation angle equal to the cutoff angle of 10° around the globe has shown that the error of this approximation is from –1.5 to 1 mm. Thus, despite the results of some investigations [Hoque and Jakowski, 2008a] in most cases the error resulting from the thin layer approximation does not exceed 2 mm (see also [Bassiri and Hajj, 1992, 1993]).

The thin layer approximation has been used in most studies of the second-order ionospheric effects In the earliest studies, the thin layer approximation was exploited to examine the influence of the second-order correction on the assessment of the accuracy in determining GPS station positions [Kedar et al., 2003; Fritsche et al., 2005; Hernandez-Pajares et al., 2007] and on the adjustment of parameters of satellite orbits, parameters of Earth rotation, and tropospheric delay [Fritsche et al., 2005; Hernandez-Pajares et al., 2007]. The analysis conducted in [Kedar et al., 2003; Fritsche et al., 2005] helped to reveal systematic northward and southward drifts of positions of receivers, taking account of the second-order correction in determining these positions. The thin layer approximation allowed developing regional models of ionospheric correction for Germany [Hoque and Jakowski, 2006] and Europe [Hoque and Jakowski, 2008b]. More recently, the formulas put forward in [Hoque and Jakowski, 2006; Hoque and Jakowski, 2008b] for estimating the second-order ionospheric error were extrapolated to the globe [Zhang et al., 2011].

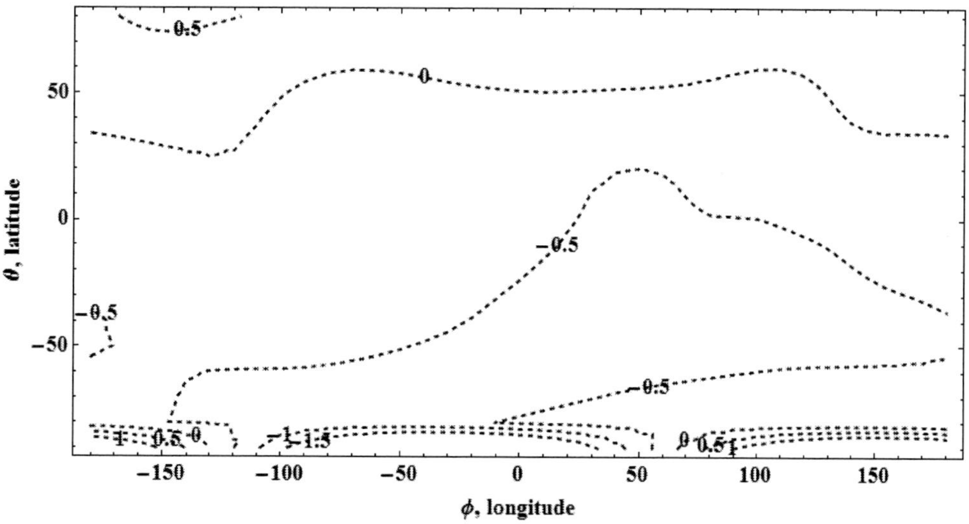

Figure 3. The dependence of the error in the thin layer approximation on GNSS observer coordinates for the elevation angle of 30° and the azimuth of 180°. On the isolines are the values of the error given in millimeters.

The third-order corrections are determined by the third term in refraction index expansion (16) and by I_{3bn} (25) associated with the ray bending. This ray bending may be caused by both the background ionospheric irregularity and random irregularities; moreover, because of the non-linear relationship between this correction and $N(\mathbf{r})$, its average depends not only on the background electron density $\bar{N}(\mathbf{r})$, but on fluctuations as well. On the other hand, the variance of the third-order error depends not only on the fluctuations $\tilde{N}(\mathbf{r})$, but also on the background irregularity $\bar{N}(\mathbf{r})$.

Given (3), it is easy to find the GNSS signal group path from phase path (19):

$$P_g(f) = D + \Delta D_{1g}^{(1)}(f) + 2\Delta D_{2g}^{(1)}(f) + 3\Delta D_{3g}^{(1)}(f) \tag{31}$$

2.3. Eliminating Ionospheric Errors in Multifrequency GNSS

Eq. (20)–(26) show that corrections in the GO approximation take the form

$$\Delta D_{ig}^{(1)}(f) = d_i / f^{i+1}, \tag{32}$$

where d_i is frequency independent. Such frequency dependence (32) enables us to eliminate ionospheric corrections in multifrequency measurements.

Leaving only a first-order correction in (19), in dual-frequency measurements we have

$$\varphi_1 = \varphi_g(f_1) = D - 40.3 I_{1g} / f_1^2, \tag{33}$$

$$\varphi_2 = \varphi_g(f_2) = D - 40.3 I_{1g} / f_2^2. \tag{34}$$

This system allows determining distance D and TEC I_{1g} from linear combinations of phase measurements

$$D \approx D^{(2)} = \frac{\varphi_1 f_1^2 - \varphi_2 f_2^2}{f_1^2 - f_2^2}, \tag{35}$$

$$I^{(2)} = \frac{f_1^2 f_2^2}{40.3} \frac{\varphi_1 - \varphi_2}{f_1^2 - f_2^2}. \tag{36}$$

Expressions (35), (36) are usually called ionosphere- and geometry-free combinations respectively.

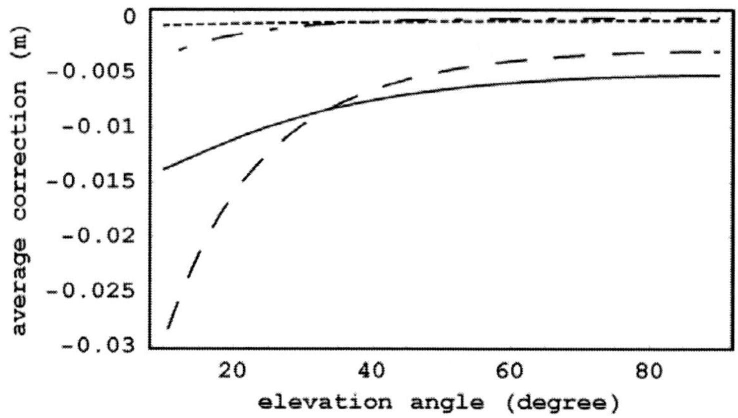

Figure 4. Elevation angle dependences of the average corrections of dual-frequency GNSS caused by the geomagnetic field (solid line), background irregularity (dash-dotted line), random irregularities (dashed line), and by the square term of the refractive index (dotted line).

Ionosphere-free combination (35) is used to remove the first-order ionospheric error in dual-frequency GNSS receivers. To find the dual-frequency measurement error (termed residual error), we substitute complete phase expression (19) in view of (20)–(26) instead of φ_1 and φ_2 in (35). The result is [Kim and Tinin, 2011]:

$$D^{(2)}(f_1, f_2) \approx D - \Delta D_{2g}^{(2)}(f_1, f_2) - \Delta D_{3g}^{(2)}(f_1, f_2), \tag{37}$$

where the second and third-order corrections in dual-frequency reception (or corresponding residual errors of opposite signs) are defined as

$$\Delta D_{2g}^{(2)}(f_1, f_2) = -40.3 I_{2g} / \left[(f_1 + f_2) f_1 f_2 \right], \tag{38}$$

$$\Delta D_{3g}^{(2)}(f_1, f_2) = -(40.3)^2 I_{3g} (f_1 f_2)^{-2}. \tag{39}$$

Figure 4 (see also [Kim and Tinin, 2007a]) presents elevation angle dependences of the average corrections of dual-frequency GPS ($f_1 = 1{,}575.42 \times 10^6\, Hz$, $f_2 = 1{,}227.6 \times 10^6\, Hz$) caused by the geomagnetic field (solid line), background irregularity (dash-dotted line), random irregularities (dashed line), and by the square term of refractive index (dotted line).

In the simulation, hereafter for the background ionosphere we take the model employed above for calculating the results shown in Figure 3. In calculating the geomagnetic field effect from formula (29) we put $f_H(z_m) = 10^6\, Hz$ and took the angle $\theta(z_m) = 0$. This matches maximum of this effect. Random irregularities were given by the isotropic Gauss function of correlation

$$\Psi_N(\mathbf{r}_1,\mathbf{r}_2)=\Psi_N(\mathbf{r}_-,\mathbf{r}_+)=\sigma_N^2(\mathbf{r}_+)\exp\left\{-\frac{\mathbf{r}_-^2}{2l_N^2(\mathbf{r}_+)}\right\}, \tag{40}$$

where $\mathbf{r}_+=(\mathbf{r}_1+\mathbf{r}_2)/2$, $\mathbf{r}_-=\mathbf{r}_1-\mathbf{r}_2$, $l_N=1km$, and the root-mean-square deviation $\sigma_N(\mathbf{r}_+)$ of electron density fluctuations was set proportional to the electron density of the background ionosphere

$$\sigma_N(\mathbf{r}_+)=\sigma_0\bar{N}(\mathbf{r}_+) \tag{41}$$

In the simulation, we chose $\sigma_0=0.08$.

Figure 4 shows that the third-order correction contribution associated with the third (square) term in the refraction index expansion (16) is within 2 mm (see also [Bassiri and Hajj, 1992, 1993; Hoque and Jakowski, 2008; Kim and Tinin, 2007a]). Thus, hereafter we will ignore it and relate the third-order correction only to the ray deviation from a straight line, setting in (22) and (39)

$$I_{3g}\approx I_{3bn} \tag{42}$$

It is also obvious that the third-order effects associated with the ray bending in random irregularities may exceed second-order effects. At the same time, ray bending effects in the background ionosphere do not normally exceed second-order ones.

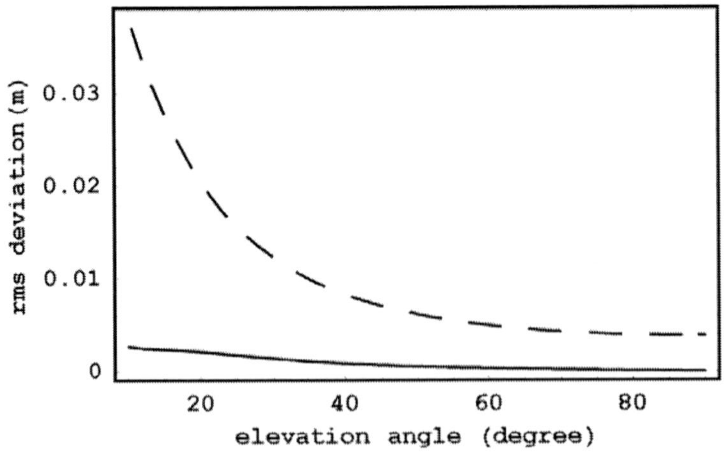

Figure 5. The elevation angle dependencies of root-mean-square deviations of the third-order, dual-frequency reception errors for the contribution σ_{DR} (dashed line) that does not take into account background inhomogeneity and the contribution σ_{DB} (solid line) taking into account the regular refraction. The parameters are the same as in Figure 4.

Figure 5 illustrates the results of the numerical simulation of the root-mean-square deviation of the dual-frequency reception error. The parameters are the same as for Figure 4. As the dual-frequency reception (35) eliminates first-order error, the error fluctuations here are totally determined by the third-order error fluctuations. The dashed line depicts the elevation angle dependence for the component σ_{DR} disregarding the background irregularity; the solid line shows the σ_{DB} calculation results demonstrating the relation between the fluctuations and the refraction in the background ionosphere. As is seen from Figure 5, $\sigma_{DB} < \sigma_{DR}$, i.e. the presence of the background irregularity slightly affects fluctuations of this error. Besides, the calculation results imply that the fluctuations may be fairly strong.

The considerable increase in the accuracy of GNSS measurements when using dual-frequency measurements poses the question about further improvement of the accuracy by increasing number of frequencies. In fact, much as for the dual-frequency measurements, we can make up a system of three equations from abbreviated equations (19), using measurements in three frequencies. Bassiri [1990], Wang et al [2005], Gherm et al [2006], and Xu [2007] suggested eliminating thus the first- and second-order corrections at a time. However, the third-order effects are ignored here which, as was shown by the above-mentioned simulation results, may stand comparison with the second-order effects. On the other hand, the above relation (29) between the second-order correction and TEC I_{1g} (and, consequently, the first-order correction) allows us to remove both first- and second-order effects already in dual-frequency measurements. For this purpose, neglecting the third-order correction in (19) and accounting for (29), we get

$$\varphi_g(f) \approx D - 40.3 I_{1g} f^{-2} - 40.3 f_H(z_m)\cos\theta(z_m) f^{-3} I_{1g} \approx D - 40.3 I_{1g} f_{ef}^{-2}, \tag{43}$$

where

$$f_{ef} = f - f_H(z_m)\cos\theta(z_m)/2. \tag{44}$$

Now, solving equation system derived from the measurements (43) at two frequencies, we obtain (see also [Kim and Tinin, 2011; Tinin and Konetskaya, 2011]):

$$D \approx D^{(2)}(f_1, f_2) = \frac{\varphi_1 f_{ef1}^2 - \varphi_2 f_{ef2}^2}{f_{ef1}^2 - f_{ef2}^2}, \tag{45}$$

$$f_{efi} = f_i - f_H(z_m)\cos\theta(z_m)/2, \quad i = 1,2.$$

Eq. (45) differs from linear combination (35) in a small change of weighting factors in measuring phases φ_1 and φ_2 at frequencies f_1 and f_2 respectively. This modification allows eliminating both first- and second-order corrections with the aid of linear combination (45).

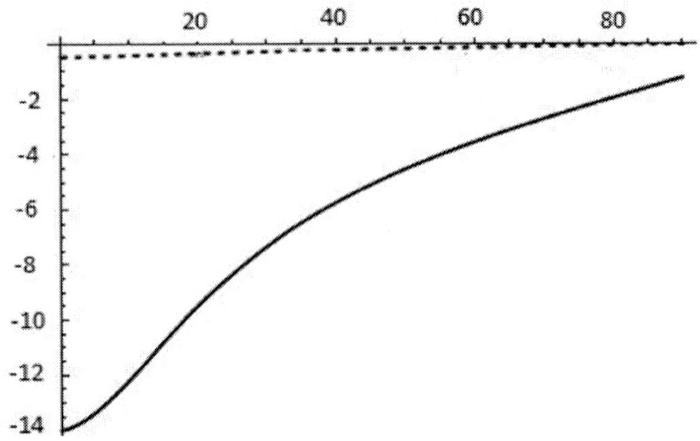

Figure 6. The satellite elevation angle dependence of the residual error for common combination of dual-frequency measurements (35) (solid line) and its modification (45) (dashed line). The receiver's coordinates are 0° N, 45° E; the azimuth is 180°; $f_1 = 1,575.42 \times 10^6\, Hz$ and $f_2 = 1,227.6 \times 10^6\, Hz$.

Figure 6 presents the results of simulating the satellite elevation angle dependence of the residual error for GPS frequencies $f_1 = 1,575.42 \times 10^6\, Hz$ and $f_2 = 1,227.6 \times 10^6\, Hz$. Here we employed common ionosphere-free combination of dual-frequency phase path measurements (35) (solid line) and its modification (45) (dashed line). For the geomagnetic field model we took the International Geomagnetic Reference Field (IGRF). The receiver was at the point with coordinates 0° N, 45° E, and the azimuth was set equal 180°. Figure 6 shows that when we pass on from combination (35) to its modification (45) the residual error at low elevation angles decreases by 95%; at nearly 90° elevation angles (vertical sounding), by 97%. The simulation results demonstrate that the high degree of removing residual error (95–97%) keeps constant all over the Earth's surface.

Now formula (19) may be written without the second-order correction as follows

$$\varphi_g(f) \approx D - 40.3 I_{1g} f_{ef}^{-2} - (40.3)^2 I_{3g} f_{ef}^{-4} \tag{46}$$

Measurements $\varphi_g(f)$ at three frequencies enable us to make up a system of three observation equations. Solving this system yields [Tinin et al., 2004; Kim et al., 2005; Kim and Tinin, 2007b, 2011; Tinin and Konetskaya, 2011]

$$D \approx D^{(3)}(f_1, f_2, f_3) = a_1 f_{ef1}^2 \varphi_1 + a_2 f_{ef2}^2 \varphi_2 + a_3 f_{ef3}^2 \varphi_3 \quad , \tag{47}$$

where

$$a_1 = f_{ef1}^2 / \left[\left(f_{ef1}^2 - f_{ef2}^2 \right)\left(f_{ef1}^2 - f_{ef3}^2 \right) \right], \; a_2 = f_{ef2}^2 / \left[\left(f_{ef2}^2 - f_{ef1}^2 \right)\left(f_{ef2}^2 - f_{ef3}^2 \right) \right],$$
$$a_3 = f_{ef3}^2 / \left[\left(f_{ef3}^2 - f_{ef1}^2 \right)\left(f_{ef3}^2 - f_{ef2}^2 \right) \right]. \tag{48}$$

Thus, in three-frequency reception the representation of phase measurements in the form of (46) makes it possible to eliminate the second- and third-order errors at a time. Note that the phase representation in the form of (46) was obtained in the geometrical optics approximation.

3. Diffraction Effects in GNSS Measurements

The spectrum of ionospheric irregularities is rather wide. The smallest sizes l_m can usually be seen in the range of 100m [Yeh and Liu, 1982]. As is known [Kravtsov and Orlov, 1990], the condition of validity of the geometrical optics (GO) approximation requires that sizes of irregularities exceed the Fresnel radius. The Fresnel radius for GNSS frequencies at heights of 300 km is estimated to be 200–700 m depending on the elevation angle (see, for example, [Kim and Tinin, 2011]). So, there are irregularities in the ionosphere for which conditions of the GO approximation are not fulfilled. To take into account the influence of diffraction by these irregularities on the multifrequency correction, described in the previous section, requires us to turn to the wave approach. To do this, we employ the method of smooth perturbations (the Rytov approximation) [Ishimaru, 1978; Rytov et al., 1989; Tatarskii, 1971]. Since we used the second approximation for the eikonal in the GO approximation, we will exploit the second-order Rytov approximation here as well.

3.1. Ionospheric Wave Phase in View of Diffraction Effects

Let the z axis in our coordinate system $\{x, y, z\}$ be directed along propagation of an incident plane wave. The quasi-longitudinal approximation, as was said above, is good to describe GNSS signal propagation. In this approximation, Maxwell equations for the combination of x and y - components of the radio-wave electric field $\mathbf{E}$

$$U_{+,-} = E_x \mp iE_y \tag{49}$$

may yield [Kim and Tinin, 2009b]:

$$\Delta U_{+,-} + k^2 n_{+,-}^2(\mathbf{r}) U_{+,-} = 0\,. \tag{50}$$

If in the expression for the refractive index $n_{+,-}^2$ we keep only those components that, as was evident from the GO approach, are the most essential (see (14) - (15)), it will become:

$$n_{+,-}^2(\mathbf{r}) = 1 - X(\mathbf{r})\left[1 \pm Y_L(\mathbf{r})\right] = 1 + \varepsilon_{1+,-}\,, \tag{51}$$

where $Y_L(\mathbf{r}) = Y\cos\theta$, $\varepsilon_{1+,-} = -X(\mathbf{r})\left[1 \pm Y_L(\mathbf{r})\right]$.

As opposed to the usual approach [Ginzburg, 1970; Budden, 1985], the waves $U_{+,-}$ considered are respectively right- and left-handed polarized waves rather than ordinary and extraordinary ones.

According to the index of refraction extraordinary wave will be the one for which $\pm Y_L(\mathbf{r}) > 0$.

For the right-handed polarized GNSS signal we will analyze only $U_+ = U$.

By applying the second Rytov approximation [Tatarskii, 1971] to solve Eq.(50), we obtain [Kim and Tinin, 2009a,b]:

$$U(\mathbf{r}) = \exp[\Phi(\mathbf{r})], \tag{52}$$

where the complex phase $\Phi(\mathbf{r})$ is given by

$$\Phi(\mathbf{r}) = \Phi_0(\mathbf{r}) + \Phi_1(\mathbf{r}) + \Phi_2(\mathbf{r}) + \ldots, \tag{53}$$

$$\Phi_0 = ik[z - z_0] = ikD, \tag{54}$$

$$\Phi_1 = \frac{k^2}{4\pi}\int_{z_0}^{z} dz' \int_{-\infty}^{\infty}\int_{-\infty}^{\infty} d^2\rho' \frac{\varepsilon_{1+}(\boldsymbol{\rho}', z')}{z - z'} \exp\left\{\frac{ik}{2}\frac{(\boldsymbol{\rho} - \boldsymbol{\rho}')^2}{z - z'}\right\}, \tag{55}$$

$$\Phi_2 = \frac{1}{4\pi}\int_{z_0}^{z_t} dz' \int_{-\infty}^{\infty}\int_{-\infty}^{\infty} d^2\rho' \frac{(\nabla\Phi_1)^2}{z - z'} \exp\left\{\frac{ik}{2}\frac{(\boldsymbol{\rho} - \boldsymbol{\rho}')^2}{z - z'}\right\}. \tag{56}$$

Given (51), complex phase (53) yields for the phase path [Kim and Tinin, 2009a,b]:

$$\begin{aligned} \varphi_d(f) &\equiv \operatorname{Im}\Phi / k \approx D - \Delta D_{1d}^{(1)}(f) - \Delta D_{2d}^{(1)}(f) - \Delta D_{3d}^{(1)}(f) \\ &= D - 40.3 f^{-2} I_{1d}(f) - 40.3 f^{-3} I_{2d}(f) - (40.3)^2 f^{-4} I_{3d}(f) \end{aligned}, \tag{57}$$

where

$$I_{1d}(f) = \int_{z_0}^{z_t} dz' \int_{-\infty}^{\infty}\int_{-\infty}^{\infty} d^2\kappa N_\kappa(\boldsymbol{\kappa}, z') \cos\left\{\frac{\kappa^2}{2}\frac{(z_t - z')}{k}\right\}, \tag{58}$$

$$I_{2d}(f) = \int_{z_0}^{z_t} dz' \int_{-\infty}^{\infty}\int_{-\infty}^{\infty} d^2\kappa N_\kappa(\boldsymbol{\kappa}, z') f_H(z') \cos\theta(z') \cos\left\{\frac{\kappa^2}{2}\frac{(z_t - z')}{k}\right\}, \tag{59}$$

$$I_{3d}(f) = -\frac{1}{2}\int_{z_0}^{z} dz' \int_{z_0}^{z'}\int_{z_0}^{z'} dz_1 dz_2 \int_{-\infty}^{\infty}\int_{-\infty}^{\infty} d^2\kappa_1 \int_{-\infty}^{\infty}\int_{-\infty}^{\infty} d^2\kappa_2 \boldsymbol{\kappa}_1\boldsymbol{\kappa}_2 N_\kappa(\boldsymbol{\kappa}_1, z_1) N_\kappa(\boldsymbol{\kappa}_2, z_2) \times$$
$$\cos\{S(\boldsymbol{\kappa}_1, \boldsymbol{\kappa}_2, z', z_1, z_2)\},$$

(60)

$$S(\boldsymbol{\kappa}_1, \boldsymbol{\kappa}_2, z', z_1, z_2, k) = \frac{\kappa_1^2}{2}\frac{(z_t - z_1)}{k} + \frac{\kappa_2^2}{2}\frac{(z_t - z_2)}{k} + \boldsymbol{\kappa}_1\boldsymbol{\kappa}_2 \frac{(z_t - z')}{k}, \tag{61}$$

$$N_\kappa(\boldsymbol{\kappa}, z) = \frac{1}{(2\pi)^2}\int_{-\infty}^{\infty}\int_{-\infty}^{\infty} d^2\rho\, N(\boldsymbol{\rho}, z) \exp\{-i\kappa\boldsymbol{\rho}\} \tag{62}$$

is a two-dimensional spectrum of the electron density of ionospheric plasma. When sizes of the irregularities l_N satisfy inequality

$$l_N >> r_{Fr} = \sqrt{\lambda(z - z')}, \tag{63}$$

formulas (57) – (60) transform to corresponding GO approximation ones; therefore $\Delta D_{1d}^{(1)}(f)$, $\Delta D_{2d}^{(1)}(f)$, $\Delta D_{3d}^{(1)}(f)$ may be called diffraction variants of the first-, second- and third-order corrections respectively. Scales of the background irregularity of ionospheric plasma $\bar{N}(\mathbf{r})$ are much larger than the Fresnel radius r_{Fr}. Thus, taking into account the weak influence of the ionospheric irregularities on the second-order effects, we can use GO formulas (28)–(29) for calculating these effects. Besides, within the framework of the thin layer approximation we can include the second-order correction in the background (geometrical optics) component of the first-order correction, as was done in deriving (46), and obtain:

$$\varphi_d(f) = D - \frac{40.3}{f_{ef}^2}\bar{I}_{1g} - \frac{40.3}{f_{ef}^2}\tilde{I}_{1d}(f) - \left(\frac{40.3}{f_{ef}^2}\right)^2 I_{3d}(f), \tag{64}$$

where

$$\tilde{I}_{1d}(f) = \int_{z_0}^{z_t} dz' \int_{-\infty}^{\infty}\int_{-\infty}^{\infty} d^2\kappa \tilde{N}_\kappa(\boldsymbol{\kappa}, z') \cos\left\{\frac{\kappa^2}{2}\frac{(z_t - z')}{k}\right\} \tag{65}$$

is a random component of (58).

3.2. Diffraction Effects and Multifrequency Measurements

Taking account of the diffraction effects has yielded that part of the corrections in the observation equation (64) is not of the form of d_i / f^{i+1} like corrections in the GO approximation and can not be eliminated in multifrequency measurements. This is true for the components of the first- and third-order corrections associated with the irregularities with sizes less than the Fresnel radius.

Substituting (64) in dual-frequency combination (45) gives [Kim and Tinin, 2011]

$$D_d^{(2)}(f_1, f_2) \approx D - \Delta D_{1d}^{(2)}(f_1, f_2) - \Delta D_{3d}^{(2)}(f_1, f_2), \tag{66}$$

where

$$\Delta D_{1d}^{(2)}\left(f_1, f_2\right) = -40.3\left[\tilde{I}_{1d}(f_1) - \tilde{I}_{1d}(f_2)\right] / \left(f_{ef1}^2 - f_{ef2}^2\right) \tag{67}$$

is the first-order correction and

$$\Delta D_{3d}^{(2)}\left(f_1, f_2\right) = -\left(40.3\right)^2\left[I_{3d}(f_1) f_{ef1}^{-2} - I_{3d}(f_2) f_{ef2}^{-2}\right] / \left(f_{ef1}^2 - f_{ef2}^2\right) \tag{68}$$

is the third-order one.

Substituting (64) in triple-frequency combination (47) yields [Kim and Tinin, 2011]:

$$D_d^{(3)}(f_1, f_2, f_3) = D - \Delta D_{1d}^{(3)}(f_1, f_2, f_3) - \Delta D_{3d}^{(3)}(f_1, f_2, f_3), \tag{69}$$

where corrections of triple-frequency measurements of first $\Delta D_{1d}^{(3)}(f_1, f_2, f_3)$ and third $\Delta D_{3d}^{(3)}(f_1, f_2, f_3)$ orders respectively have the form of

$$\Delta D_{1d}^{(3)}(f_1, f_2, f_3) = \Delta \tilde{D}_{1d}^{(3)}(f_1, f_2, f_3) = 40.3\left[a_1 \tilde{I}_{3d}(f_1) + a_2 \tilde{I}_{3d}(f_2) + a_3 \tilde{I}_{3d}(f_3)\right], \tag{70}$$

$$\Delta D_{3d}^{(3)}(f_1, f_2, f_3) = \left(40.3\right)^2\left[a_1 I_{3d}(f_1) f_{ef1}^{-2} + a_2 I_{3d}(f_2) f_{ef2}^{-2} + a_3 I_{3d}(f_3) f_{ef3}^{-2}\right]. \tag{71}$$

For large-scale irregularities for which condition (63) is fulfilled, (58), (60) are independent of frequency and

$$\Delta D_{1d}^{(2)}(f_1, f_2) = \Delta D_{1d}^{(3)}(f_1, f_2, f_3) = \Delta D_{3d}^{(3)}(f_1, f_2, f_3) = 0\,. \tag{72}$$

Note that the average first-order correction is also zero when allowing for diffraction effects in both dual-frequency and triple-frequency measurements. Hence the average correction here can be determined only through ray bending in random irregularities. Figs. 7–10 present simulation results (see also [Kim and Tinin, 2011; Tinin and Konetskaya, 2011]) of the elevation angle dependence of the average and root-mean-square deviations of corrections in multifrequency measurements for $f_1 = 1{,}575.42 \times 10^6\, Hz$, $f_2 = 1{,}227.6 \times 10^6\, Hz$, $f_3 = 1{,}176.45 \times 10^6\, Hz$. For the background ionosphere we took model (30) with parameters used above for simulating the ionospheric corrections in the GO approximation. However, Gauss model (40) is not appropriate enough to describe random irregularities in the turbulent multiscale ionospheric plasma. A more suitable model for the ionospheric turbulence spectrum

$$\Phi_N(\mathbf{\kappa}, \mathbf{r}_+) = (2\pi)^{-3} \int_{-\infty}^{\infty}\int_{-\infty}^{\infty}\int_{-\infty}^{\infty} d^3 r_- \Psi_N(\mathbf{r}_-, \mathbf{r}_+) \exp\{-i\mathbf{\kappa}\mathbf{r}_-\} \quad (73)$$

is the model (see, for example, [Rytov et al., 1989]):

$$\Phi_N(\mathbf{\kappa}, z) = \frac{\sigma_N^2(z)\Gamma(p/2)\kappa_0^{p-3}}{\pi^{3/2}\Gamma((p-3)/2)} \left(\kappa_0^2 + \kappa^2\right)^{-p/2} \exp\left[-\kappa^2 / \kappa_m^2\right], \quad (74)$$

where $\sigma_N^2(z)$ is variance of electron density fluctuations; $\kappa_0 = 2\pi / L_0$, $\kappa_m = 2\pi / l_m$; L_0 and l_m are outer and inner scales respectively.

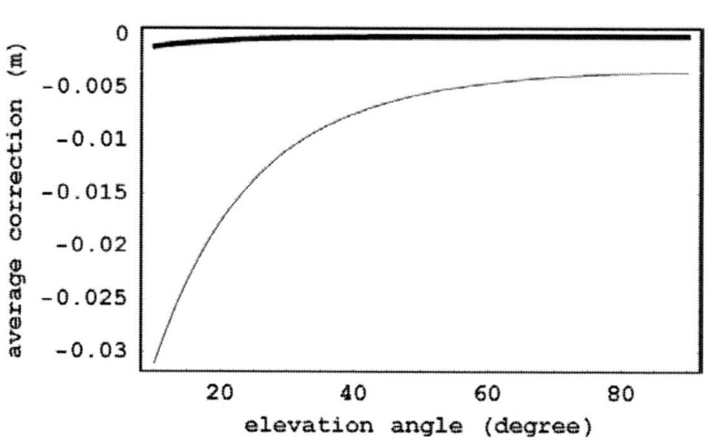

Figure 7. The elevation angle dependencies of average corrections of dual-frequency (thin line), and triple-frequency (thick line) measurements. As opposed to the results shown in Figure 4 and 5, here we take the spectrum (74) for the turbulent ionosphere with outer and inner scales equal to $L_0 = 20km$, $l_m = 1km$ respectively.

In the simulation, as in Subsection 2.3, we employed assumption (41) about proportionality of the variance of electron density fluctuations to the square of its average with the proportionality factor $\sigma_0^2 = 0.0064$.The outer scale L_0 was taken equal 20 km; the index, p=11/3.

Figure 7 presents results of the calculation of average corrections in dual-frequency (thin line) and triple-frequency (thick line) measurements when the inner scale $l_m = 1km$. The accuracy improves considerably when we switch from dual-frequency to triple-frequency measurements. Figure 8 for the same parameters as in Figure 7 shows results of calculations of root-mean-square deviations of the first-order (solid lines) and third-order (dashed lines) corrections for dual-frequency (thin lines) and triple-frequency (thick lines) measurements.

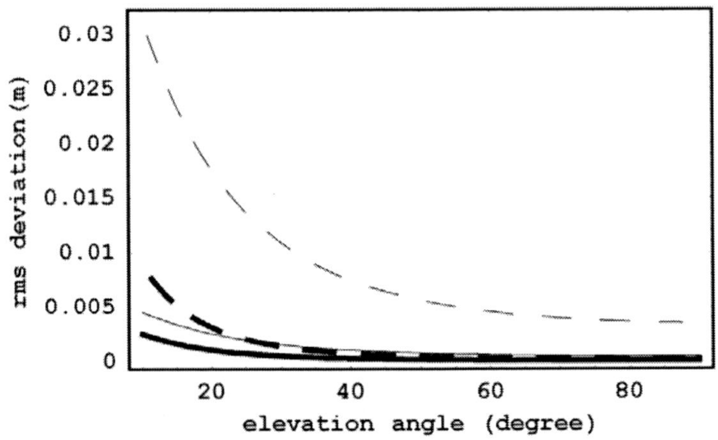

Figure 8. The elevation angle dependencies of root-mean-square deviations of dual-frequency (thin line), and triple-frequency (thick line) measurements for the same parameters as in Figure 7.

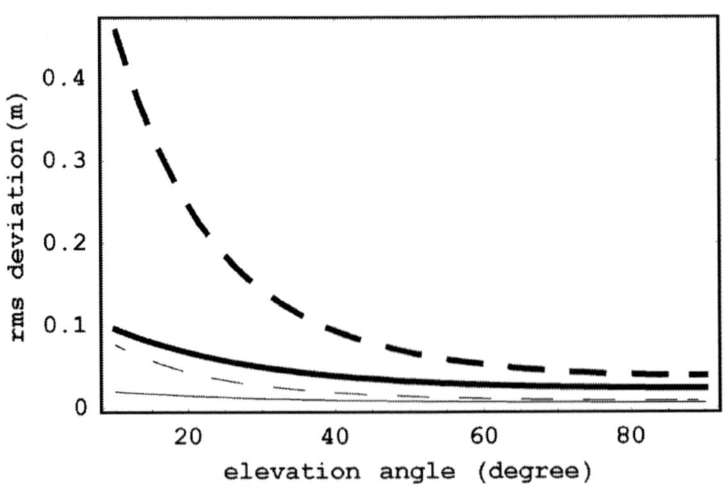

Figure 9. The same as in Figure 8 for $l_m = 0.1km$.

The calculation results imply that when we pass on to triple-frequency measurements the average correction decreases along with its fluctuations. Now let us examine a situation when the inner scale is less than the Fresnel radius. For this we take $l_m = 0.1km$. Results of the simulation of the average correction differ slightly from those presented in Figure 7 and therefore are left out here. However, as is obvious from Figure 9, at $l_m < r_{Fr}$ the change to triple-frequency measurements in this case increases error fluctuations.

CONCLUSION

The analysis of the results of studying the ionospheric corrections and the possibilities of eliminating them has demonstrated that the first- and second-order corrections associated with radio wave propagation in the large-scale background ionosphere can be eliminated in dual-frequency measurements by modifying the well-known ionosphere-free measurement combination. The numerical simulation of the third-order effects caused by the ray bending in random irregularities has highlighted an essential contribution they make to the error of GNSS measurements similar to that of the second-order effects. The third-order effects associated with large-scale irregularities can be removed by triple-frequency GNSS measurements.

Ionospheric effects can largely be eliminated in multifrequency GNSS if their contributions to phase and time measurements are represented as series in inverse powers of frequency. This representation is a result of the application of the geometrical optics (GO) approximation to the description of GNSS signal propagation in the inhomogeneous ionosphere. However, the presence of irregularities with scales less than the Fresnel radius in the turbulent ionosphere necessitates using a diffraction model for describing GNSS signals instead of the GO approximation. In this model, the phase path can not be represented as series in inverse powers of frequency, and therefore Fresnel-type diffraction effects do not allow us to totally eliminate residual errors of all orders, including the first-order correction, at any number of frequencies in use. Nevertheless, the numerical simulation has revealed that the average of a residual error decreases when we pass on from dual- to triple-frequency measurements independently of the relation between the Fresnel radius r_{Fr} and the minimum scale of the ionospheric turbulence l_m. But this relation essentially governs the behaviour of fluctuations of the residual error during the transition from dual-frequency GNSS to triple-frequency ones. If the Fresnel radius is less than l_m, then during this transition both the average of residual errors and their variances decrease. Otherwise, diffraction may increase the random (noise) component of the residual error with increasing number of GNSS measurement frequencies.

Diffraction effects can be eliminated by the spatial-temporal processing used in radio occultation observations [e.g. Gorbunov et al., 2000; Melbourne, 2004; Liou et al., 2010], in which two approaches are generally employed. The first approach is founded on ideas of the method of canonical operator (Maslov method) to account for the caustic diffraction effects associated with the appearance of multipath effects and caustics during wave propagation in an inhomogeneous medium (e.g. [Gorbunov, 2002]). In this method [Maslov and Fedoryuk,

1976; Kravtsov and Orlov, 1999], the Fourier transform is exploited to represent the desired field in the vicinity of a caustic as the superposition (integral) of plane waves which can be described with the GO approximation because their caustics emerge in another place. This approach requires prior information on the ray structure in the inhomogeneous medium considered. In the diagnostics of the ionosphere and troposphere through radio occultation observations, the horizontal inhomogeneity of medium is usually ignored, and the ray structure can be approximately determined a priori. In the randomly inhomogeneous medium, the ray pattern is very changeable that complicates the use of the Maslov method. Moreover, results of this approach are in rather poor agreement with those of the Rytov approximation [Tinin, 1998], and consequently it is difficult to use them for taking account of Fresnel diffraction effects appearing during wave propagation in a medium with irregularities of sizes less than the Fresnel radius.

Another approach (see, for example, [Rytov et al., 1989]) is based on the Helmholtz–Kirchhoff integral theorem (Green formula) which, if we know the Green function for a region between two planes, makes it possible to convert the field from one plane to another. In this case, it is often suggested that the plane wave coming from a celestial source is incident on a relatively thin ionospheric layer and, when leaving the layer, acquires only a phase modulation. During further propagation to the Earth's surface, Fresnel diffraction effects and caustics with multipath effects, which can be described by the Helmholtz–Kirchhoff theorem, appear. This phase (virtual) screen model is widely used in direct problems of wave propagation in the randomly inhomogeneous ionosphere (see, for instance, [Rino, 2011]).

The processing of radio occultation observations by the back propagation technique [Mortensen and Hoeg, 1998; Melbourne, 2004] relies on the same approach, but it is used for inverse conversion from the measured field values to the values at the output from the layer (screen) where there are no diffraction effects due to the small thickness of the layer. This way allows eliminating diffraction effects in GNSS measurements. However, there are problems of phase screen positioning [Barabanenkov et al., 1971].

Kravtsov and Tinin [2000], Tinin et al. [2002], Kravtsov and Tinin [2003] put forward the double weighted Fourier transform (DWFT) method to describe waves in inhomogeneous media. It exploits the Fourier transform over coordinates of the source and receiver. This method in the first approximation of the perturbation theory correlates well with results of the phase screen and Rytov approximation, and consequently accounts for both Fresnel- and caustic-type diffraction effects. Using the inverse DWFT, we can get a super Fresnel resolution in tomographic problems [Tinin et al., 2002; Kravtsov and Tinin, 2003; Tinin and Kravtsov, 2008; Kravtsov et al., 2009; Kravtsov et al., 2010] and decrease amplitude fluctuations in communication and diagnostic problems [Tinin, 2010; Tinin and Kim, 2011]. But GNSS measurements have no possibility of spatial processing of a signal in two planes. Hence, to apply DWFT to eliminate diffraction effects in multifrequency GNSS, we should pass on from the double transform to the single one. Tinin [2012] has modified DWFT in this way and obtained the generalization of the phase screen method for the case of an extended inhomogeneous medium remote from the source and the observer. To use this generalization for improving multifrequency GNSS accuracy, as was shown above (see Subsection 3.1), requires employing the second approximation of the perturbation theory to calculate third-order effects. This necessitates developing the DWFT method accounting for the effects of random ionospheric irregularities in the second approximation of the perturbation theory.

So, technological advance in state-of-the-art GNSS sets researchers in radio wave propagation new tasks considered until recently to be only academic, devoid of practical application.

ACKNOWLEDGMENTS

We are very grateful to O.A. Kulish for her assistance in preparing the English version of the manuscript. This work was performed with the partial support of the Russian Foundation for Basic Research (grant 12-02-00249) and the Ministry of Education and Science of the Russian Federation (Project N 8388).

REFERENCES

Afraimovich, E. L.; Ishin, A. B.; Tinin, M. V.; Yasyukevich, Yu. V.; Jin, S. G. *Adv. Space Res.* 2011, 47, 1674-1680.

Alber, C.; Ware, R.; Rocken, C.; Solheim, F. *Geophys. Res. Lett.* 1997, 24, 1859–1862.

Ashmanets, V. I.; Vodyannikov, V. V.; Troitsky, B. V. *J. Atmos. Terr. Phys.* 1996, 58, 1161-1163.

Barabanenkov, Yu. N.; Kravtsov, Yu. A.; Rytov, S. M.; Tatarskii, V. I. *Soviet Physics Uspekhi.* 1971, 13, 551-575.

Bassiri, S. *TDA Prog. Rep.* 1990, 42-103, 14-20.

Bassiri, S; Hajj, G. A. *TDA Prog. Rep.* 1992, 42-110, 92-103.

Bassiri, S; Hajj, G. A. *Manuscr. Geod.* 1993, 18, 280-289.

Bonnor, N. *J. Nav.* 2012, 65, 1–14.

Brunner, F. K.; Gu, M. *Manuscr. Geod.* 1991, 16, 205-214.

Budden, K. G. The propagation of radio waves; Cambridge University Press: Cambridge, 1985.

Datta-Barua, S.; Walter, T.; Blanch, J.; Enge, P. *Radio Sci.* 2008, 43, RS5010, doi:10.1029/2007RS003772.

El-Rabbany, A. Introduction to GPS: the Global Positioning System; Artech House: Norwood, 2002.

Fritsche, M.; Dietrich, R.; Knöfel, C.; Rülke, A.; Vey, S.; Rothacher, M.; Steigenberger, P. *Geophys. Res. Lett.* 2005, 32, L23311, doi:10.1029/2005GL024342.

French, G.; T. Understanding the GPS. An Introduction to the Global Positioning System; GeoResearch: Bethesda, 1996.

Gershman, B. N.; Erukhimov, L. M.; Yashin, Yu. Ya. Wave Phenomena in the Ionosphere and Cosmic Plasma; Nauka: Moscow, 1984 [in Russian].

Gherm, V.; Novitsky, R.; Zernov, N.; Strangeways, H. J.; Ioannides, R. T. 2nd Workshop Radio systems and ionospheric effects. Rennes, 3-7 October 2006, Paper11.

Ginzburg, V. L. The Propagation of Electromagnetic Waves in Plasmas; Gordon and Breach: New York, 1970.

Gu, M.; Brunner, F. K. *Manuscr. Geod.* 1990, 15, 357-361.

Gorbunov, M. E. *Radio Sci.* 2002, 37, 1076, doi:10.1029/2000RS002592.

Gorbunov, M. E.; Gurvich, A. S.; Kornblueh, L. *Radio Sci.* 2000, 35, 1025-1034.

Grewal, M, S.; Weill, L. R.; Andrews, A. P. Global Positioning Systems, Inertial Navigation, and Integration; Wiley: New York, 2007.

Harper, N. Server-Side GPS and Assisted-GPS in Java; Artech House: Norwood, 2010.

Hernández-Pajares, M.; Juan, J. M.; Sanz, J.; Orús, R. *J. Geophys. Res.* 2007, 112, B08417, doi:10.1029/2006JB004707.

Hooijberg, M. Geometrical Geodesy; Springer-Verlag: Berlin, 2008.

Hoque, M. M.; Jakowski, N. *J. Geod.* 2006, 81, 259–268.

Hoque, M. M.; Jakowski, N. *Radio Sci.* 2008a, 43, RS5008, doi:10.1029/2007RS003817.

Hoque, M. M.; Jakowski, N. *GPS Solut.* 2008b, 12, 87-97.

Ishimaru, A. Wave Propagation and Scattering in Random Media. Vol 2. Multiple Scattering, Turbulence, Rough Surfaces and Remote Sensing; Academic Press: New York, 1978.

Ivanov, V. B.; Gephan, G. D.; Gorbachev, O. A. *J. Atmos. Sol. Terr. Phys*. 2011, 73, 1703–1707.

Kedar, S.; Hajj, G. A.; Wilson, B. D.; Heflin, M. B. *Geophys. Res. Lett.* 2003, 30, 1829, doi:10.1029/2003GL017639.

Kennedy, M. The Global Positioning System and GIS: An Introduction; Taylor and Francis: London, 2002.

Kim, B. C.; Kolesnik, S. N.; Tinin, M. V. Proc. XXVIII URSI GA, New Delhi, India, October 2005, GF1b.8.

Kim, B. C.; Tinin, M. V. 7th International Symposium on Antennas, Propagation and EM Theory Proceedings, Oct. 26-29, 2006 Guilin, China, Vol. I, 433-436.

Kim, B. C.; Tinin, M. V. *J. Geod.* 2007a, 81, 189–199.

Kim, B. C.; Tinin, M. V. *Geomag. Aeron*. 2007b, 47, 238–243.

Kim, B. C.; Tinin, M. V. *Wave. Random Complex*. 2009a, 19, 284–304.

Kim, B. C.; Tinin, M. V. *J. Atmos. Sol. Terr. Phys.* 2009b, 71,1967–1973.

Kim, B. C.; Tinin, M. V. *J. Geod.* 2011, 85, 159–169.

Klobuchar, J. A.; Kunches, J. M. *GPS Solut*. 2003, 7, 55-58.

Kravtsov, Yu. A.; Orlov, Yu. I. Geometrical Optics of Inhomogeneous Media; Springer-Verlag: New York, 1990.

Kravtsov, Yu. A.; Orlov, Yu. I. Caustics, Catastrophes and Wave Fields; Springer-Verlag: New York, 1999.

Kravtsov, Yu. A.; Tinin, M. V. *Radio Sci*. 2000, 35, 1315-1322.

Kravtsov, Yu. A.; Tinin, M. V. *Cosmic Res*. 2003, 41, 357–358.

Kravtsov, Yu. A.; Tinin, M. V.; Kulizhsky, A. V. *Fusion Eng. Des.* 2009, 84, 1113-1115.

Kravtsov, Yu. A.; Tinin, M. V.; Knizhnin, S. I. *J. Commun. Techn. Electron.* 2011, 56, 831–837.

Kravtsov, Y. A.; Tinin, M. V.; Knizhin, S. I. 20th URSI International Symposium on Electromagnetic Theory, Berlin, Germany, August 16-19, 2010, Symposium Proceedings, *IEEE Xplore*, 601-604.

Kunches, J. M.; Klobuchar, J. A. *GPS Solut.* 2001, 4, 52-54.

Kursinski, E.; Hajj, G.; Schofield, J.; Linfield, R.; Hardy, K. *J. Geophys. Res.* vol. 102, 23429–23465, 1997.

Langley, R. B. In GPS for Geodesy; Kleusberg, A,; Teunissen, P. J. G. Ed; Lecture Note in Earth Science, 60; Springer-Verlag: Berlin, 1998a; pp 141-174.

Langley, R. B. In GPS for Geodesy; Kleusberg, A,; Teunissen, P. J. G. Ed; Lecture Note in Earth Science, 60; Springer-Verlag: Berlin, 1998b;103-140.

Liou, Y. A.; Pavelyev, A. G.; Matyugov, S. S.; Yakovlev, O. I.; Wickert, J. Radio Occultation Method for Remote Sensing of the Atmosphere and Ionosphere; In-Theh: Vukovar, Croatia, 2010.

Maslov, V. P. Fedoryuk, M. V. Quasi-Classical Approximation for Quantum Mechanics Equations; Nauka: Moscow, 1976 (in Russian).

Melbourne, W. G. Radio Occultations Using Earth Satellites: A Wave Theory Treatment; Jet Propulsion Laboratory: California, 2004.

Moore, R. C.; Morton, Y. T. *Radio Sci.* 2011, 46, RS1008, doi:10.1029/2010RS004380..

Mortensen, M. D.; Hoeg P. *Geophys. Res. Lett.* 1998, 25, 2441-2444.

Moritz, H. Österr Zeitschrift für Vermessungswesen. 1967, 25, 323–334.

de Munck, J. C. The theory of dispersion applied to electrooptical distance measurement and angle measurement. Netherlands Geodetic Commission, Publications on Geodesy, New Series 3(4), Delft, 1970.

Prasad, R.; Ruggieri, M. Applied Satellite Navigation Using GPS, GALILEO, and Augmentation Systems; Artech House: Norwood, 2005.

Rino, C. L. The Theory of Scintillation with Applications in Remote Sensing; John Wiley and Sons: New Jersey, 2011.

Rytov, S. M.; Kravtsov, Yu. A.; Tatarskii, V. I. Introduction to Statistical Radiophysics, vol. 4, Wave Propagation through Random Media; Springer-Verlag: New York, 1989.

Samama, N. Global positioning: technologies and performance; John Wiley and Sons: New Jersey, 2008.

Seeber, G. Satellite geodesy; Walter de Gruyter: Berlin, 2003.

Schaal, R. E.; Larocca, A. P. C. *GPS Solut.* 2002, 5, 15–18.

Strangeways, H. J.; Ioannides, R. T. *Acta Geod. Geoph.* 2002, 37, 281-292.

Tatarskii, V. I. The effect of a turbulent atmosphere on wave propagation; National Technical Information Service: Springfield, 1971.

Tinin, M. V. *Waves Random Media.* 1998, 8, 329–338.

Tinin, M. V. 20th URSI International Symposium on Electromagnetic Theory, Berlin, Germany, August 16-19, 2010, Symposium Proceedings, *IEEE Xplore*, 483-486.

Tinin, M. V. *Radiophys. Quant. Electr.* 2012 (submitted).

Tinin, M. V.; Kolesnik, S. N.; Kravtsov, Yu. A. *Proc. SPIE.* 2002, 4678, 605-609.

Tinin, M. V.; Kim, B. C.; Kolesnik, S. N.; Kravstov, Yu. A. Proc. ISAP'04. 2004, Sendai, Japan, 1105-1108.

Tinin, M. V.; Kim, B. C. *Wave. Random Complex.* 2011, 21, 645-656.

Tinin, M. V.; Konetskaya, E. V. URSIGASS-2011 Symposium Proceedings, Istanbul, Turkey, August 13-20, 2011, *IEEE Xplore*, doi: 10.1109/URSIGASS.2011.6050862.

Tinin, M. V.; Kravtsov, Yu. A. *Plasma Phys. Controlled Fusion.* 2008, 50, doi: 10.1088/0741-3335/50/3/035010.

Tsui, J. B. Fundamentals of global positioning system receivers: a software approach; John Wiley and Sons: New Jersey, 2005.

Wang, Z.; Wu, Y.; Zhang, K.; Meng, Y. J. *Global Positioning Syst.* 2005, 4, 291–295.

Williams, D. C. *Surv. Rev.* 1975, 23, 166–172.

Xu, G. GPS Theory, Algorithms and Applications; Springer: New York, 2007.

Yeh, K. C.; Liu, C. H. *Proc. IEEE.* 1982, 70, 324-360.

In: Recent Advances in Satellite Research and Development ISBN: 978-1-62417-443-8
Editors: Samuel Gardiner and Kenneth P. Olsen

Chapter 5

MODELING POACHED ELEPHANTS HOTSPOTS IN TSAVO EAST NATIONAL PARK, KENYA

Daniel Muteti Kyale[1], Shadrack Ngene[2], Shadrack Muya[3] and John Maingi[4,*]

[1]Kenya Wildlife Service, Tsavo Conservation Area, Voi, Kenya
[2]Kenya Wildlife Service, Elephant Programme, Nairobi, Kenya
[3]Department of Zoology, Jomo Kenyatta University of Agriculture and Technology, Nairobi, Kenya
[4]Geography Department, Miami University, Oxford, Ohio, US

ABSTRACT

This study investigated the spatial-temporal patterns of elephant poaching in Tsavo East National Park to establish and map the density of poached elephants as well as model elephant poaching hotspots. Data on poached elephants for the period between 10 October 1989 and 2 July 2005 was acquired from the Kenya Wildlife Service elephant mortality database at the headquarters. The results indicate a higher concentration of overall poaching mainly in the central and northern parts of the park. Both dry and wet season elephant poaching ellipses indicate that seasonal elephant poaching exhibited similar spatial orientation and their centers of distribution lay approximately in the central part of the park. Kernel density analysis depicted dry season poaching hotspots on the Tiva and Galana rivers in the northwest and central-western parts of the park. Sites with moderate dry season elephant poaching included the southeastern tip of the park on Voi River and the central-eastern part. There were two main wet season poaching hotspots occurring along the lower part of Galana River across the park. Moderate wet season poaching sites were spread along Tiva River to the north of the park through the central parts of the park, to the southeastern tip on Voi River. The risk to elephant poaching map for wet season indicated areas at highest risk as occurring mainly in the central-east,

* Daniel Muteti Kyale: Kenya Wildlife Service, Tsavo Conservation Area, P.O. Box 14, Voi, Kenya. Shadrack Ngene: Kenya Wildlife Service, Elephant Programme, P.o. Box 40241, Nairobi 00100, Kenya. Shadrack Muya: Department of Zoology, Jomo Kenyatta University of Agriculture and Technology, P.O. Box 62,000 - 00200 Nairobi, Kenya. John Maingi: Geography Department, Miami University, 33A Shideler Hall, 501 East High Street, Oxford, Ohio, US.

south of Sala gate toward the southeastern edge of park, and in the northeastern parts of the park. The area of highest risk to elephant poaching during the dry season occurred along the Tiva River, central parts of the Galana and Voi rivers. Elephant risk to poaching during the dry season indicated areas of highest risk as concentrated along the main rivers, where water is available throughout the year. An overall risk to poaching map indicated areas with the highest risk to elephant poaching occurred in the central part of the Tiva River and the central-eastern parts of the Galana River. In order to provide adequate security of elephants throughout the year, there is need to re-deploy wildlife policing resources (equipment and personnel) to high risk areas as well as have temporally security mobile units in the open plains during the wet season. There is also need to open more security patrol roads on the eastern side of Sangayaya and Ithumba areas.

Keywords: Elephant poaching, hotspot, Tsavo East, National Park, mortality

INTRODUCTION

Elephants are megaherbivores and play various important roles in any ecosystem they occupy. They are known to aid seed dispersal in Asian forests (Pragya, 2011). They also influence trophic structure in African ungulate communities (Fritz et al., 2002), thus functioning as key stone species and umbrella species. They are also a flagship species that drive many conservation programs. Thus their survival is very important in sustaining healthy ecological units. However, ultimate survival of any wildlife species is a function of several factors that may be classified as either biophysical and/or anthropogenic. Biophysical factors include access to water, roads, slope, elevation, terrain ruggedness and vegetation. Anthropogenic factors include distance to roads (Khaemba and Stein, 2000), proximity to human habitation (Mpanduji et al., 2002; Douglas-Hamilton, 1998) and overexploitation (Diamond 1984 and, Harper 1960). These factors influence the distribution and habitat preferences for different animal species in their natural habitats (Musiega and Kazadi, 2004).

As such, an animal will select a habitat that enhances its chances of survival and reproduction (Kyale, 2006). Unfortunately, not all the preferred habitat actually enhances an animal's survival and reproduction success. Some become mortality zones if they are located where wildlife and human interests are bound to clash (Delibes et al., 2001), or at worse are targeted by poachers. Various studies have been conducted on spatial aspects of elephants including population size and distribution (Barnes et al., 1995; Omullo et al., 1998, Ngene, 2010), movement, home range estimation and habitat preferences (Douglas-Hamilton, 1998; Babaasa, 2000; Mpanduji et al., 2002, Ngene, 2010), diet (Ben-Shahar, 1999; De Boer et al., 2000), and human-elephant conflicts (Hoare, 2000; Smith and Kasiki, 2000; Sitati et al., 2003). However, little has been done on modeling spatial temporal human-caused elephant mortality, yet this could be helpful predicting poaching likelihood and deploying of necessary preventive measures. Nevertheless, available literature does provides invaluable information about biophysical and human factors that tend to influence elephant distribution and by extension may influence patterns of elephant mortality (Verlindern and Gavor, 1998; Ben-Shahar, 1999; Smith and Kasiki, 2000; Khaemba and Stein, 2000; Mpanduji et al., 2002; Ngene et al., 2009). Demeke and Bekele (2000) established that elephant carcasses at Mago National Park in Ethiopia occurred mainly along riverine and forest areas during the dry season. A study on population and distribution of elephants in the central sector of Virunga National Park in the Democratic Republic of Congo (Ruggiero, 1990) established that elephants exhibited compressed distribution preferring to aggregate in localized places, a characteristic reminiscent of elephant populations that had experienced heavy of poaching (Ruggiero, 1990). Pilgram and Western (1986) and Leader-Williams et al (1990) hypothesize that poaching occurs in remote areas with abundant elephants logically because remote areas are difficult for patrol rangers to access due to inadequate routes. Inaccessibility to remote areas can thus afford poachers a chance to poach elephants without their activities being detected by patrol rangers (Kyale, 2006). Poaching, law enforcement and ivory trade have been identified as among the top priority issues affecting the African elephant in the elephant range states (AFESG, 1997; WWF, 1997). Protection of elephants in their natural ranges has been estimated to cost approximately US$ 200 per km^2 annually (Cumming et al., 1990).

Kenya, like many other poor Sub-Saharan African countries struggling to feed their people and keep their economies afloat is unable to meet this financial obligation and is thus unlikely to allocate more funds towards wildlife conservation. Jachmann and Billiouw (1997) estimated that there should be a minimum of one park ranger for every 24 km^2 of wildlife reserve if effective patrolling and policing is to be realized. Kenya Wildlife Service, like most wildlife departments in other African countries, is understaffed with approximately one ranger per 100 km^2 of wildlife reserve. It is therefore important to explore strategies that involve more efficient use of the limited available resources.

By assessing spatial and temporal patterns of elephant mortality, important insights on characteristics of Tsavo East National Park (TENP) elephants were obtained that in turn can help guide effective implementation of anti-poaching policy. TENP was chosen as a preferred study area because of four reasons. First, it is the largest park in the country (KWS, 2003). Second, it has both the highest concentration of elephants (Economist, 2002; Hammer, 1993). Third, it has the highest incidences of elephant poaching reported 1990s (Robinson, 2000). Four, TENP is a predominantly semi-arid bushland with only a small area of the park developed and open for tourism. Unfavorably hot climate, poor accessibility, and the large size of the park, make patrolling difficult and more challenging with currently available resources (Kioko, 2002). Within the context of previous studies, and the available management options, we used GIS to describe patterns of elephant mortality attributed to poaching in TENP and addressed the following questions: are there any clear temporal and spatial distribution patterns of poached elephants in TENP?; are there areas in TENP that could be considered to be at a greater risk of elephant poaching than others?; and, in light of the temporal-spatial poaching patterns evidence, what are the appropriate conservation strategies for the management of elephants in TENP?

MATERIALS AND METHODS

Study Area

The Tsavo East National Park was gazetted in 1948, and is one of the oldest pioneer national parks in Kenya which included Nairobi gazetted in 1946, Amboseli gazeted in 1947 and Mount Kenya gazetted in 1949 (Akama, 1998; Smith and Kasiki, 2000). It covers approximately 13,747 km^2, accounting for about 40% of the total area covered by parks in Kenya (KWS, 2003; Smith and Kasiki, 2000). Elevation within the park increases westwards from 150 m in the eastern park boundary to 1,200 m in the western boundary (Tolvanen, 2004). Rainfall occurs twice per year with the long rains being experienced from late March and May, and the short rains occurring from late October and December. The dry season is experienced from January to February and June to September. Like elevation, rainfall in the park increases from about 250 mm in the eastern part to about 450 mm in the western part (Smith and Kasiki, 2000). The main source of permanent water in the park is the Galana River. This river is formed by the union of Athi and Tsavo rivers respectively. Seasonal rivers and other seasonal sources of water in the park include: the Tiva and Voi rivers, the Aruba dam and a few scattered ponds and swamps (Figure 1).

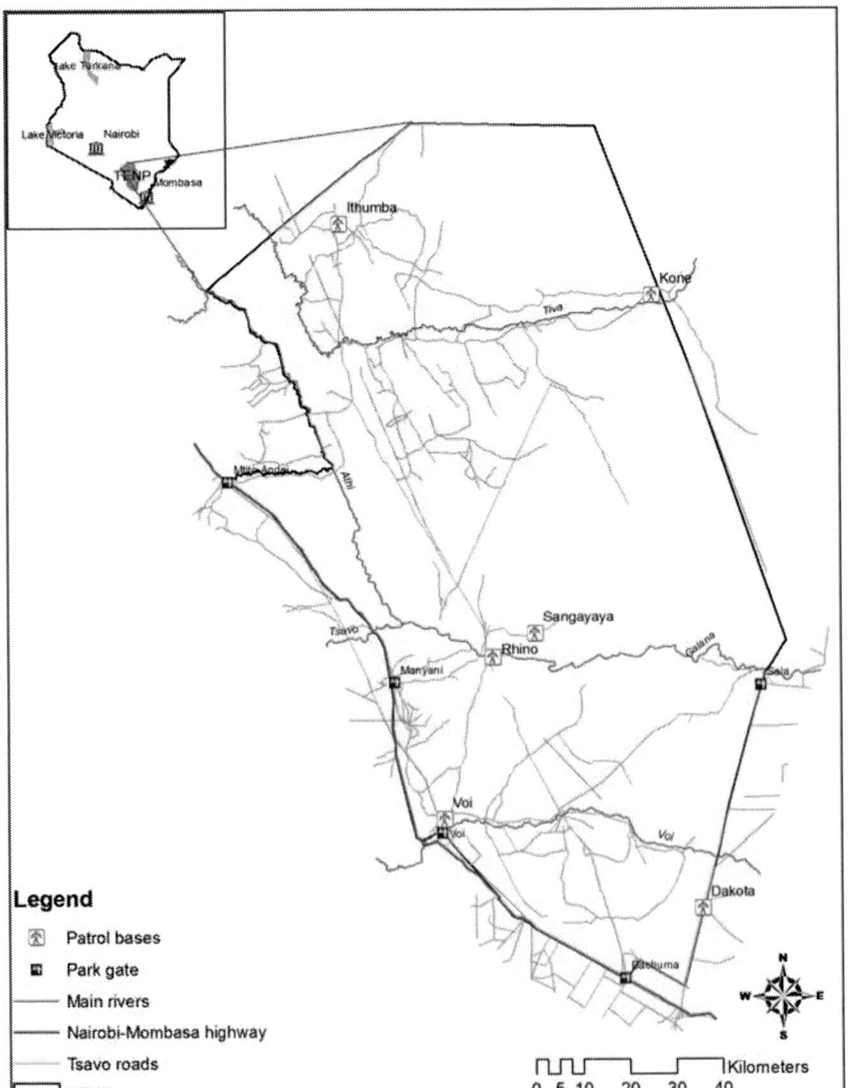

Source: Muteti, 2006.

Figure 1. Location of Tsavo East National Park in Southern Kenya.

The vegetation in the TENP is mainly composed of bushland, grassland, and woodland savannah (McKnight, 2000). A common vegetation community is the semi-arid *Acacia* and *Commiphora* woodlands. The woodlands also consists of *Premna, Bauhinia* and *Sericocomorpsis* shrubs, scattered *Delonix eleta* and *Melia volkensii* trees, which are interspersed with open plains (McKnight, 2000). Riverine vegetation dominated by *Acacia elatior, Hyphaene compressa*, and *Suaeda monoica* occurs along the rivers (McKnight, 2000). The western part of the park has more dense vegetation than the eastern part, which corresponds to a decreasing rainfall gradient (Smith and Kasiki, 2000).

Data on Elephant Mortality

Seven years data on elephant mortality in TENP was obtained from the Kenya Wildlife Service (KWS) elephant mortality database at the KWS headquarters and covered the period between 10 October 1989 and 2 July 2005. The mortality data were then entered in an Excel spreadsheet. Seventy five (wet season = 40 and dry season = 35) records were used in the analysis. Each record had the following fields: X,Y-axis coordinates (in UTM), date of mortality, cause of mortality, and name of location where mortality occurred. The data was further converted into dBASE IV format and shapefiles for dataset created using ArcGIS 9.2 (ESRI, 2006) and associated attribute data (Figure 2). Each dataset category was analyzed to discern its spatial patterns as described by Mitchell (2009).

A 10-km buffer of TENP was generated and used to clip elephant mortality data to the buffer area (Figure 2).

The 10-km buffer was sufficient to reduce edge effects during point pattern analysis (Kyale, 2006). The spatial extent of all subsequent data sets generated for this study was based on the 10-km buffer.

GIS and Remote Sensing Data Layers

The GIS and remote sensing data sets used were: lands cover types, roads, rivers, waterholes, elevation, slope, patrol bases, park gates, and park boundary. A map of vegetation cover types of the study area was developed from four Landsat ETM+ images obtained during the same season (Two images, each for January and March 2000). Landsat ETM+ images were preferred because of the reasonably good spatial resolution and widespread success in land cover mapping for natural resources projects (McCarthy e*t al*., 2005; Muller et al. 1999; Cherrill et al., 1994). The four images were prepared and the image classified as described by Kyale (2006) and implemented using ERDAS software (Leica Geosystems, 2003). Nine vegetation cover classes were adopted (i.e., forest, woodland, bushland, open bushland, herbaceous cover, grassland, active agriculture, barren land, and water; Wijgaarden, 1985). MODIS Normalized Difference Vegetation Index (NDVI) images, generated from 32-day 500m composites were also used as outlined by Kyale (2006) as they are good indicator of the spatial and temporal variability in food resources (water and forage) for the elephants (Eklundh, 1998; Tappan et al., 1992; Moreau et al., 2003). The NDVI composites were also useful for broadly identifying different vegetation types through interpretation of seasonal changes in vegetation phenology (Moleele et al., 2001).

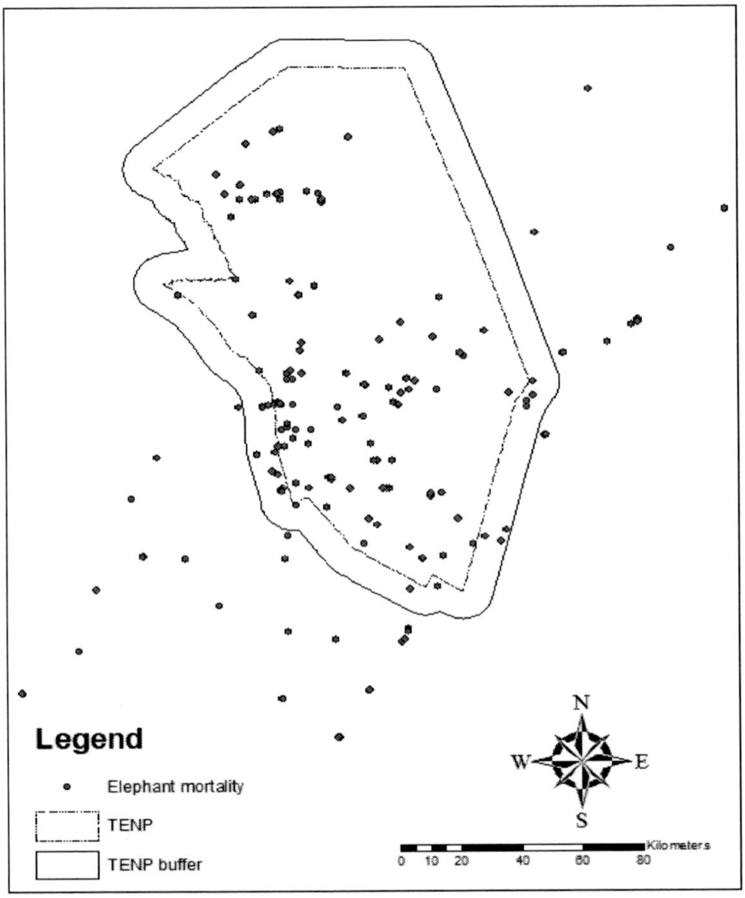

Figure 2. A point map showing the distribution of elephant carcasses in Tsavo East National Park and a 10 km buffer (1990 to 2005).

Classification accuracy assessment was performed on the final map (Leica Geosystems, 2003). An error matrix was constructed and producer's, user's accuracy, overall accuracy, and Kappa statistics calculated (Leica Geosystems, 2003). The overall classification accuracy was over 85% as recommended by Anderson et al., (1976).

A 90m DEM compiled from the Shuttle Radar Topographic Mission (SRTM) was used to obtain elevation grid and slope of the study area. The other GIS data layers sets were acquired as described by Kyale (2006) and Kyale et al. (2011; Figure 3).

Data Analysis

ArcGIS 9.2 was used to generate standard deviation ellipses associated with the mean center of spatial distribution of poaching-induced elephant mortality in order to ascertain its spatial spread and directionality. Kernel density analyses were performed to identify areas within the study area that were hotspots for elephant poaching.

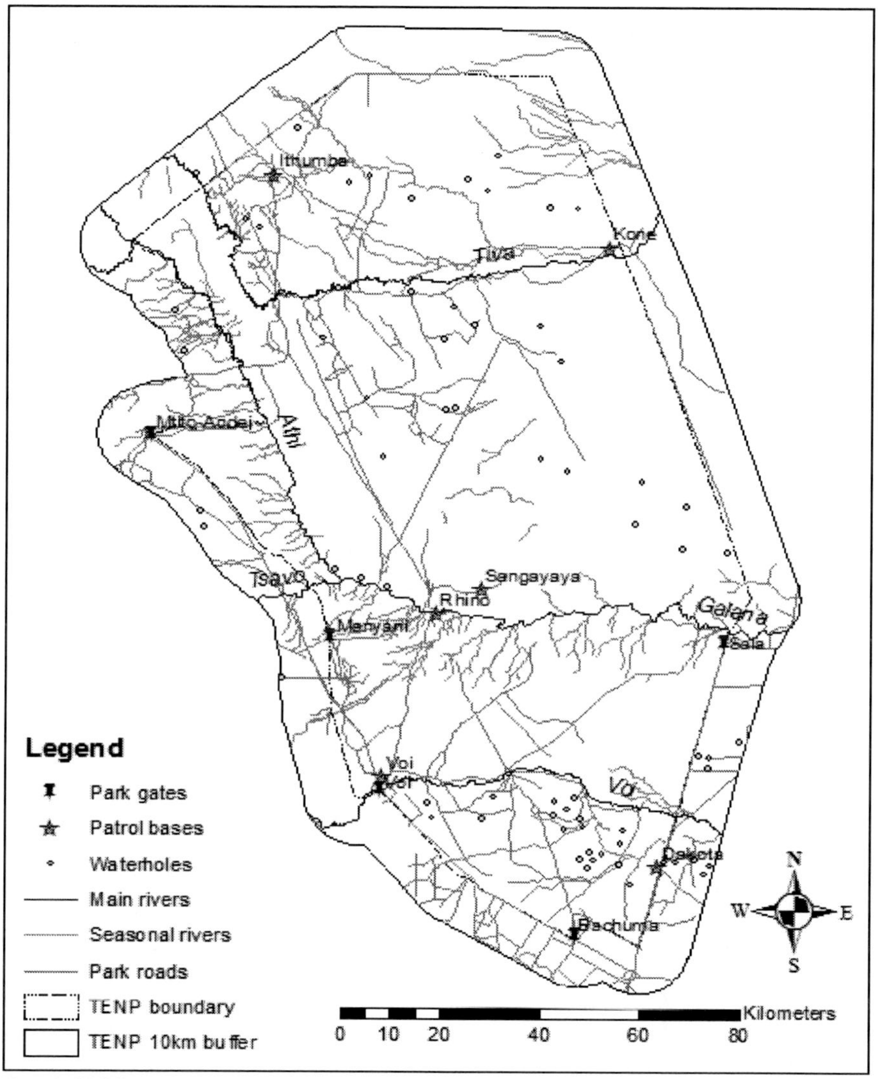

Source: Muteti, 2006.

Figure 3. GIS layers generated.

Kernel density analyses for different elephant mortalities due to poaching were performed using ArcGIS. A band width of 23.9km was selected in the analysis because it corresponds to mean home range size for TENP elephants (Leuthold and Sale, 1973; Mukeka, 2010). Female elephants in TENP have an average home range of 2400 km^2 while that for males averages at 1200 km^2 (Leuthold and Sale, 1973; Mukeka, 2010). During the dry season, however, mean elephant home range expands (Leuthold and Sale, 1973; Mukeka, 2010), and as such, a wider band width (31.5km) was used for analysis of dry season poaching.

The biophysical and anthropogenic factors in this study include: (a) distance to patrol bases (b) distance to park gates (c) distance to park roads (d) distance to park boundary (e) distance to main rivers (f) distance to seasonal rivers (g) distance to waterholes (h) elevation (i) slope (j) land cover types. ArcGIS' Spatial Analyst was used to create distance surfaces to the GIS data layers a-g above (ESRI, 2006). Additionally, elevation and slope, and land cover types at every elephant mortality location were extracted from DEM and TENP land cover map respectively (ESRI, 2006).

The biophysical and anthropogenic factors were used to generate risk to poaching surfaces for TENP using the Model Builder in ArcGIS (ESRI, 2006). The biophysical or anthropogenic factor used to create risk to poaching models was rasterized and reclassified into ten classes using equal interval method so that elephant risk to poaching would be assessed in a scale of 1 – 10. For each biophysical or anthropogenic factor classified, new values were assigned according to the relationship the biophysical or anthropogenic factor exhibited when correlated with the various elephant poaching categories analyzed. For instance, distance to main rivers, which exhibited negative correlation with dry season poaching was recorded into 10 classes. A new value of 10 was assigned to the shortest mortality location distance to the nearest main river, while a new value of 1 was assigned to the longest mortality location distance to the nearest main river. For a positive correlation, a new value of 10 was assigned to the longest mortality location distance, and a new value of 1 assigned to the shortest distance to the biophysical or anthropogenic factor being reclassified. During reclassification of land covers, new values were assigned to each land cover type in accordance to the elephant mortality density occurring in it for every elephant poaching category examined.

Results and Discussion

Standard Deviation Ellipse and Kernel Density of Poached Elephant

Standard deviation ellipse results showed a spatial distribution for overall poaching centered slightly in the northern part of the park (Figure 4a). The results indicate a higher concentration overall poaching mainly in the central and northern parts of TENP. Kernel density results depicted overall poaching hotspots as occurring mainly along the main river courses in the park (Figure 4a). Concentration of poaching hotspots along the main rivers indicates the influence of proximity to surface water on the distribution of elephants in the park because elephant drink and mud-bathe daily (Estes, 1999). The results are in agreement with Ottichilo's (1987) findings that elephant poaching was concentrated along the central part of Galana River and in the north and northwestern parts of the park.

Analysis using the standard deviation ellipse indicated a much narrower poaching pattern in the dry season compared to the wet season. Both dry and wet season elephant poaching ellipses, however, indicate that seasonal elephant poaching exhibited similar spatial orientation and their centers of distribution lay approximately in the central part of the park.

Kernel density analysis depicted dry season poaching hotspots on the Tiva and Galana rivers in the northwest and central-western parts of TENP, respectively (Figure 4b). Sites with moderate elephant poaching included the southeastern tip of the park on Voi River and the central-eastern part.

There were two main wet season poaching hotspots occurring along the lower part of Galana River across the park (Figure 4c). Moderate wet season poaching sites were spread along Tiva River to the north of TENP through the central parts of the park, to the southeastern tip on Voi River.

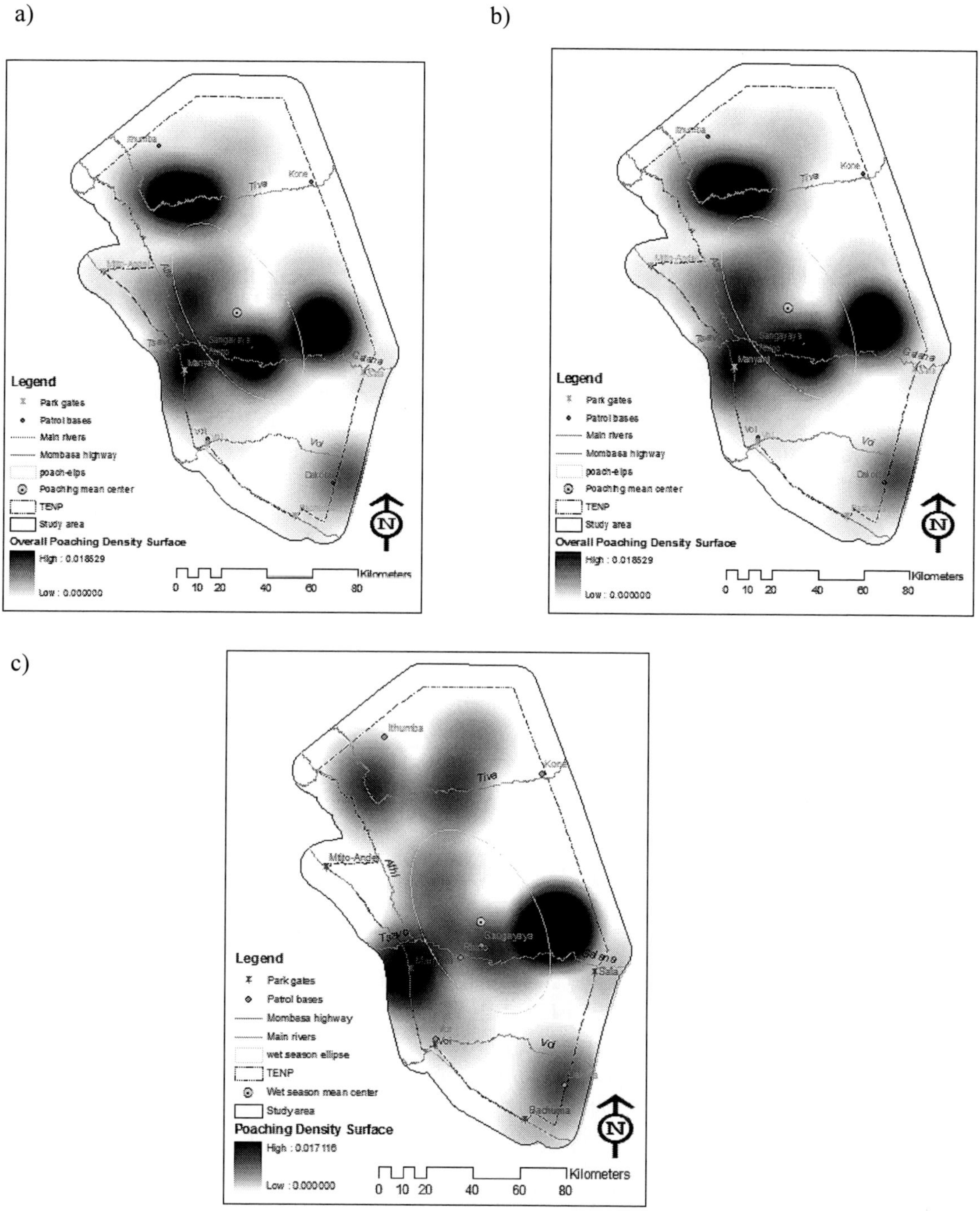

[a]Annual.
[b]Dry season.
[c]Wet season.

Figure 4. Kernel density and standard deviation ellipse results for poaching patterns in Tsavo East National Park for (a) annual, (b) dry season, and (c) wet season.

The difference in seasonal poaching mortality patterns may be attributed to changes in elephant behavior as influenced by seasonal changes in weather conditions. Although TENP elephants have been found to aggregate all year round, mixed herds have been found aggregate in larger numbers during wet seasons as bulls from bachelor herds join females for mating (Leuthold, 1976). Besides, resources are plenty of during wet season. The larger aggregates of elephant herds during wet season are distributed relative to available food resources (McKnight, 2000). Clusters of wet season elephant poaching will therefore be widely dispersed within TENP compared to dry season poaching.

During dry season, resource scarcity (food and water) limit distributions of elephants. Bachelor herds are the poachers' main target because they have the largest tusks (Poole and Thomsen, 1989). These herds have been found to increase from an average of 22 to 50 individuals in TENP during dry season (Leuthold, 1976).

Because of limited food and water resources in this season, elephant home ranges increase considerably (Leuthold and Sale 1973). The expanding home ranges largely overlap in areas with food and water resources. Elephant herds thus congregate near permanent surface water sources in the dry season.

Leuthold and Sale (1973) established that Tsavo elephants actually returned to the same grazing areas every year during dry season unlike during wet season when their distribution closely followed the rainfall patterns. The concentration of elephants near permanent water sources thus, may explain why dry season poaching was concentrated near permanent rivers.

Poaching Hotspots Maps

Three poaching hotspot maps were generated for TENP, each for wet season, dry season, and for annual elephant poaching irrespective of season. The risk to elephant poaching map for wet season indicated areas at highest risk as occurring mainly in the central-east, south of Sala gate toward the southeastern edge of park, and in the northeastern parts of TENP (Figure 5a). Ottichilo (1987) observed similar poaching concentrations in the central-eastern border of TENP near the Galana ranch. Other areas at higher risk to elephant poaching during wet season were in the central parts of Voi River, and in the immediate vicinity of the areas with the highest risk to poaching. The areas of TENP with high risk to poaching were concentrated mainly within bushland, open bushland and grassland reflecting elephant preference for more open land covers in TENP during wet season.

In addition, areas of high risk to poaching were not concentrated along the main rivers because distribution of elephant during the wet season is not limited by availability of surface water. Kasiki (1998) found that elephants were widely dispersed in areas without permanent surface water during the wet season.

The area of highest risk to elephant poaching during the dry season occurred along the Tiva River (Figure 5b). Other areas of high risk to elephant poaching in the dry season were along the central parts of the Galana and Voi rivers. In contrast to the elephant risk to poaching map for wet season, that generated for the dry season indicated areas of highest risk as concentrated along the main rivers, where water is available throughout the year. According to Kasiki (1998), elephants occur in large numbers along Galana, Voi and Tiva rivers during the dry season. However, the western portion of the Voi River had a low poaching risk because of its proximity to the Voi patrol base and park gate.

a) b) c)

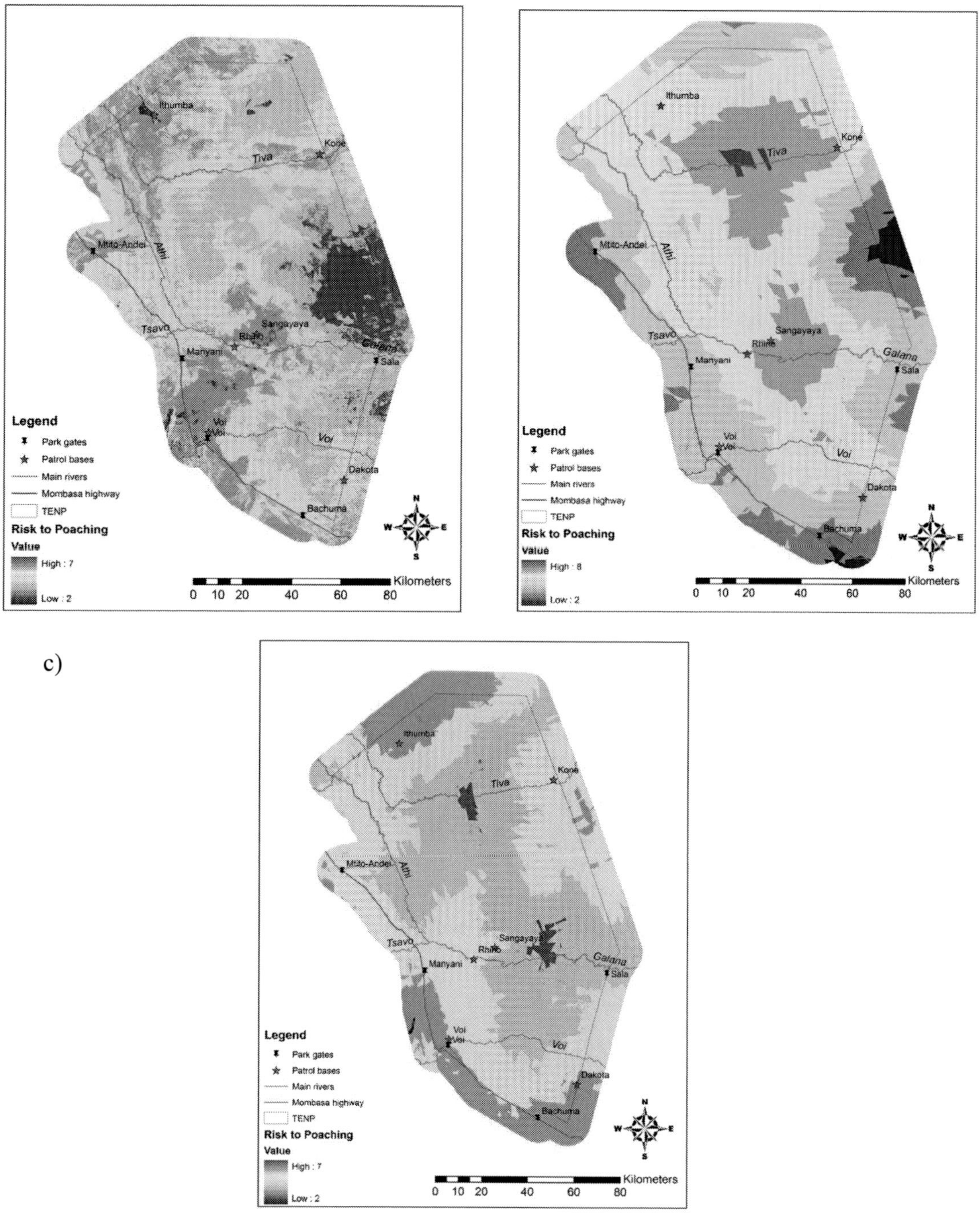

[a]Wet season.
[b]Dry season.
[c]Annual.

Figure 5. Risk to elephant poaching surface for (a) Wet Season (b) Dry Season, and (c) Annual.

Ottichilo (1987) found similar patterns with elephant poaching concentrated along the central part of Galana River and on the northern and northwestern parts of TENP. An overall risk to poaching map that did not factor the seasonality of poaching indicated areas with the highest risk to elephant poaching occurred in the central part of the Tiva River and the central-eastern parts of the Galana River (Figure 5c).

This risk to elephant poaching map had similarities to the dry season risk to elephant poaching map by showing the highest risk to poaching near sources of surface water.

It also indicated that TENP elephants were at a high risk to poaching in the more open central and eastern parts than the dense west dominated by forests and woodlands. Areas that could be considered relatively safe for elephants were found near park gates and patrol bases, except for the Sala Gate.

The location of Sala Gate close to Galana River coupled with the fact that the gate is located at the border closest to Tana River district where poachers of Somali origin often access the park may explain why elephants in this region are not safe.

Finding of this study suggest that elephants are limited in their distribution by food and water. Seasonal variations in availability of food and water resources explain observed distribution of elephants in TENP. Poachers on the other hand do not just kill elephants in any area where elephants occur in large numbers, but instead target those elephants in remote locations where poaching activities are likely to remain undetected.

Conclusion

In this paper, we map and describe areas of TENP in which elephants were at higher risk of being poached. The results show that during the dry season, risk of elephant poaching was highest in areas close to sources of permanent water. During the wet season, risk of elephant poaching was highest in areas with more open land cover types located furthest from main rivers.

In order to provide adequate security of elephants throughout the year, there is need to re-deploy wildlife policing resources (equipment and personnel) to high risk areas as well as have temporally security mobile units in the open plains during the wet season. There is also need to open more security patrol roads on the eastern side of Sangayaya and Ithumba areas.

Acknowledgments

We highly acknowledge the Miami University, Department of Geography for financially supporting Daniel Muteti Kyale during his studies at the University. For Kenya Wildlife Service (KWS), we thank all those who were involved in granting Daniel Muteti Kyale study leave during his study period. The first author acknowledges his M.Sc thesis committee members (Dr. John Maingi, Dr. Kimberly Medley, and Dr. Mary Henry), for their constructive input into the M.Sc thesis, which resulted to preparation of this paper. We also thank all the KWS staff for the assistance they provided during the first author's fieldwork in Kenya. Finally, the first author thanks the Departments of Geography and Zoology for providing financial support during his fieldwork in Tsavo East National Park, Kenya.

REFERENCES

AFESG, (1997). Review of African Elephant Conservation Priorities (Thouless, C. R, eds). Working document of the IUCN/SSC African Specialist Group, Nairobi, Kenya.

Akama, J. S. (1998). The evolution of wildlife conservation policies in Kenya. *Journal of Third world Studies,* Vol. XV(2): 103 – 117.

Anderson, J. R., Hardy, E. E., Roach, J. T., and Witmer, R. E. (1976). *A land use and land cover classification system for use with remote sensing data.* Geological Survey professional paper 964, US Government Printing Office, Washington DC, US.

Babaasa, D. (2000). Habitat Selection by Elephants in Bwindi Impenetrable National park, South Western Uganda. *African Journal of Ecology*, 38: 116-122.

Barnes, R. F. W., Blom, A., Alers, M. P. T., and Barnes, K. L. (1995). An estimate of the numbers of forest elephants in Gabon. *Journal of Tropical Ecology* 11: 27-37.

Ben-Shahar, R. (1999). Elephants and their Woodland Habitats in Northern Botswana. *Pachyderm,* 27: 101-104.

Cherrill, A. J., Lane, A. and Fuller, R. M. (1994). The use of Landsat 5 Thematic Mapper imagery in the characterization of landscape composition: a case study in northern England. *Journal of Environmental Management*, 40: 357 – 377.

De Boer, W. F., Ntumi, C. P., Correia, A. U., and Mafuca, J. M. (2000). Diet and Distribution in the Maputo Elephant Reserve, Mozambique. *African Journal of Ecology*, 38: 188-201.

Delibes, M., Goano, P. and Ferreras, P. (2001). Effects of an Attractive Sink Leading to Maladaptive habitat Selection. *American Naturalist* 158: 277-285.

Demeke, Y. and Bekele, A. (2000). Study on the Elephants of Mago National Park, Ethiopia. *Pachyderm,* 28: 32-43.

Diamond, J. (1984). "Normal "extinctions of isolated populations. In: M. H. Nitecki (ed). *Extinctions*. University of Chicago Press, Chicago.

Douglas-Hamilton, I. (1998). Tracking African elephants with a global positioning system (GPS) radio collar. *Pachyderm*, 25: 81-92.

Economist (2002). Closing in for the Kill? After a decade of quiet, ivory poaching has started again in Kenya. *Economist*, 363(8273): 76-77.

Eklundh, L. (1998). Estimating relations between AVHRR, NDVI and rainfall in East Africa at 10-day monthly time scales. *International Journal of Remote Sensing*, 19: 563-570.

Esri [Environmental Systems Research Institute] (2006) *ArcGIS: Arc Map Version 9.2*. ESRI, California, US.

Estes, R. (1999). The safari companion: Guide to watching African mammals including hoofed mammals, carnivores, and primates. Chelsea Green Publishing Company, White River Junction.

Fritz, H., Duncan, P., Gordon, I., and Illius, A. (2002). Megaherbivores influence trophic guilds structure in African ungulate communities. Oecologia, 131(4), 620-625. Springer. Retrieved from http://www. springerlink.com/Index/10.1007/s00442-002-0919-3

Hammer, J. (1993). The war to Save the Tsavo Reserve. *Newsweek* Vol. 121(15): 3-4.

Harper, C. L. (1960). Environment and Society: Human Perspectives on Environmental issues. Prentice-Hall. New Jersey.

Hoare, R. E. (2000). African Elephants and Humans in Conflict: The outlook for coexistence. *Oryx*, 34: 34-38.

Jachmann, H. and Billiouw, M. (1997). Elephant Poaching and Law Enforcement in the Central Luangwa Valley, Zambia. *Journal of Applied Ecology* 34(1): 233-244.

Kasiki, S. M. (1998). Human-elephant conflict in areas adjacent to the Tsavo National Parks, Kenya. Phd dissertation, University of Kent.

Khaemba, W. M. and Stein, A. (2000). Use of GIS for a Spatial and Temporal analysis of Kenyan Wildlife with Generalised Linear Modeling. *International Journal of Geographical Information Science,* 14 (8): 833-853.

Kioko, J. M. (2002). KWS Press Release - Family of ten Elephants Gunned Down in Tsavo East, one Poacher killed, three still on the run. *Kenya Wildlife Press Release*. Nairobi.

KWS (2003). Kenya Widldife Service website, Tsavo East National Park. http://www.kws.org/tsavoeast.htm. Accessed on 20th August 2011

Kyale, D. M. (2006). Assessing spatial and temporal patterns of human-caused elephant mortality in Tsavo East National Park, Kenya. MSc thesis, Miami University, Oxford, Ohio, US.

Kyale, D. M., Ngene, S. M., and Maingi, J. (2011). Biophysical and human factors determine the distribution of poached elephants in Tsavo East National Park, Kenya. *Pachyderm* 49: 48-60.

Leader-Williams, N., Albon, S. D. and Berry, P. S. M. (1990). Illegal Exploitation of Black Rhinoceros and Elephant Populations: Patterns of Decline, Law Enforcement and Patrol Efforts in Luangwa Valley, Zambia. *Journal of Applied Ecology* 27(3): 1055-1087.

Leica Geosystems (2003). ERDAS Field Guide™ 7th ed. GIS and Mapping, LLC. Atlanta, Georgia.

Leuthold, W. (1976). Group size in elephants of Tsavo National Park and possible factors influencing it. *Journal of Animal Ecology*, 45(2): 425 – 439.

Leuthold, W. and Sale, J. B. (1973). Movements and patterns of habitat utilization of elephants in Tsavo National Park, Kenya. *East African Wildlife Journal,* 11: 369 – 384.

McCarthy, J., Gumbricht, T. and McCarthy, T. S. (2005). Ecoregion classification in the Okavango Delta, Botswana using multitemporal remote sensing. *International Journal of Remote sensing*, 26(19): 4339 – 4357.

McKnight, B. (2000). Changes in Elephant Demography, Reproduction and Group Structure in Tsavo East National Park (1966 - 1994). *Pachyderm,* 29: 15-24.

Mitchell, A. (2009). *The ESRI guide to GIS analysis: Volume 2, spatial measurements and statistics*. ESRI Press, Ridland, California, US.

Moleele, N., Ringrose, S., Arnberg, S., Lunden, B., and Vanderpost C. (2001). Assessment of vegetation indexes useful for browse (forage) prediction in semi-arid rangelands. *International Journal of Remote Sensing*, 22(5): 741-756.

Moreau, S., Bosseno, R., Gu, X. F., and Baret, F. (2003). Assessing the biomass dynamics of Andean *bofedal* and *totora* high-protein wetland grasses from NOAA/AVHRR. *Remote Sensing of Environment*, 85: 516-529.

Mpanduji, D. G., Hofer, H., Hilderbrandt, T. B., Goeritz, F., and East, M. L. (2002). Movement of Elephants in the Selous-Niassa Wildlife Corridor, Southern Tanzania. *Pachyderm,* 33: 18-31.

Mukeka, J. 2010. Analyzing the distribution of the African elephant (*Loxodonta africana)* in Tsavo, Kenya. MSc thesis, Miami University, Oxford, Ohio, US.

Muller, S. V., Racoviteanu, A. E. and Walker, D. A. (1999) Landsat MSS-derived land-cover map of northern Alaska: Extrapolation methods and a comparison with photo-interpreted and AVHRR-derived maps. *International Journal of Remote Sensing*, 20(15 and 16): 2921 – 2946.

Musiega, D. E. and Kazadi, S. (2004). Simulating the East African wildebeest migration patterns using GIS and remote sensing. *African Journal of Ecology*, 42: 355 – 362.

Ngene, S. M. (2010). Why Elephant Roam. ITC publication #171, Enschede, Netherlands.

Ngene, S. M., Skidmore, A. K., Van Gils, H., Douglas-Hamilton, I., *and* Omondi, P. O. M. *(*2009*).* Elephant distribution around a volcanic shield dominated by a mosaic of forest-savannah (Marsabit, Kenya). *African Journal of. Ecology.* **47***,* 234–245.

Omullo, L. B., Said, M. T. and Warugte, P. W. (1998). Habitat Evaluation for Wildlife Conservation and Management Techinical Report 152. department of Resource Survey and Remote Sensing, Nairobi, Kenya.

Ottichilo, W. K. (1987). The cause of the recent heavy elephant mortality in the Tsavo ecosystem, Kenya, 1975 – 80. *Biological Conservation* 41: 279 – 289.

Poole, J. H. and Thomsel, J. B. (1989). Elephants are not beetles: implications of the ivory trade for the survival of the African elephant. *Oryx*, 23: 188 – 198.

Pragya Bhatt, Narendra M. B. Pradhan, Per Wegge (2011). Seed dispersal by megaherbivores: do Asian elephants disperse Mallotus philippinensis, a main food tree in northern India and Nepal? *Journal of Natural History* Vol. 45, Iss. 15-16,

Robinnson, S. (2000). Dying for Ivory: Even African conservationists cannot agree on how best to protect elephant populations from the threat of poachers. *Time Europe*, 155(15): 44-45.

Ruggiero, R. (1990). The effects of poaching disturbance on elephant behavior. *Pachyderm*, 13: 42 – 44.

Sitati, N. W., Walpole, M. J., Smith, J., and Leader-Williams (2003). Predicting Spatial Aspects of Human-Elephant Conflict. *Journal of Applied Ecology* 40:667-677.

Smith, R. J. and Kasiki, S. (2000). A Spatial Analysis of Human-Elephant Conflict in the Tsavo Ecosystem, Kenya. *AfESG Report*. IUCN/SSC, Gland.

Tappan, G. G., Tyler, D. J., Wehde, M. E., and More, D. G. (1992). Monitoring rangeland dynamics in Senegal with Advanced Very High Frequency Resolution Radiometer data. *Geocarto International*, 7: 87-98.

Tolvanen, R. (2004). Nature Conservation Areas in Kenya- Tsavo East and West National Parks. In: Pellikka, P. J., Ylhaisi and Clark, B. (eds). Tait Hills and Kenya- Seminar, Reports and Journal of a Field Excursion to Kenya. Expedition Reports of the Department of Geography, *University of Helsinki* 40: 59-63.

Verlinden, A. and Gavor, I. K. N. (1998). Satellite Tracking of Elephants in Northern Botswana. *African Journal of Ecology*, 36: 105-116.

Western, D. (1987). Africa's Elephants and Rhinos: Flagships in crisis. *Trees*, Vol. 2(11): 343 – 346.

Western, D. (1989). Conservation without Parks: wildlife in rural landscape. In: D. Western and M. C. Pearl. (eds). *Conservation for the twenty-first Century* (pp 158-165). Oxford University press, New York.

Wijgaarden, W. (1985). Elephants – trees- grazers: Relationship between climate, soil, vegetation and large herbivores in a semi-arid ecosystem (Tsavo, Kenya). ITC Publication #4, Enschede, Netherlands.

WWF, (1997). Conserving Africa's Elephants: current issues and priorities for action (Dublin, H. T. McShane, T. O. and Newby, J. eds). WWF International, 1196 Gland, Switzerland.

In: Recent Advances in Satellite Research and Development ISBN: 978-1-62417-443-8
Editors: Samuel Gardiner and Kenneth P. Olsen

Chapter 6

ARC-SECOND ATTITUDE STABILIZATION FOR A 35- KG MICRO ASTRONOMY SATELLITE

Takaya Inamori
The University of Tokyo,
Department of Advanced Energy, Chiba, Japan

ABSTRACT

This chapter focuses on the development of a precise attitude control system for a 35 kg small astrometry satellite. In general, conventional methods used for an attitude control in standard-sized satellites are not applicable to small satellites because of constraints of power consumption, space requirements, and mass. In addition, small satellites are more greatly affected than standard-sized satellites by attitude disturbances because of their relatively small moment of inertia. Although a magnetic disturbance torque is considered negligible in standard-sized satellites, it is dominant in small satellites. To realize precise attitude stabilization in small satellites, the satellites should use an attitude control method that satisfies the abovementioned constraints and reduces the effect of magnetic disturbance. This study proposes a new method to solve these problems. Through an example of an astronomical observation mission using small satellites, it is demonstrated that precise attitude control is feasible in small satellites.

1. INTRODUCTION

In recent years, small satellites having small mass and volume have attracted increasing interest because they can be developed to realize various space missions at low cost within a short period (Heidt et al., 2000). Some small satellites have been targeted at sophisticated missions such as remote sensing and astronomical observations. In 2003, CubeSat satellites developed at several universities including the University of Tokyo were launched (Nakamura et al., 2004). These satellites weighed around 1 kg and had a volume of 10 cm × 10 cm × 10 cm, and they were developed using commercial off-the-shelf (COTS) materials; they were used by universities worldwide to perform practical space science missions. After the success of these satellites, nano- and micro-satellites were proposed for more sophisticated objectives such as remote sensing and astronomical observations. Examples of these missions include Pico-satellite for Remote sensing and Innovative Space Missions (PRISM) and Nano-Japan Astrometry Satellite Mission for INfrared Exploration (Nano-JASMINE). PRISM is a remote sensing nano-satellite weighing 8.5 kg that was developed at the University of Tokyo and launched in 2009 (Nagai et al., 2005). Its objective was to capture images with 30 m resolution, for which its attitude had to be stabilized to an accuracy of 0.001 rad/s (Inamori et al., 2011(b)). Nano-JASMINE is an astrometry nano-satellite that was developed at the Intelligent Space Systems Laboratory (ISSL), University of Tokyo, in cooperation with the National Observatory of Japan (NAOJ). Its objective was to accurately estimate the positions of stars and to update star catalogues, for which its attitude, too, had to be stabilized to a high accuracy. To satisfy the mission requirements, the spin rate of the satellite was controlled to an accuracy of 4×10^{-7} rad/s, which was difficult to achieve using conventional sensors and actuators (Inamori et al., 2011(c)). Figure 1 shows the relationship between satellite mass and the requirement for attitude stabilization. In the case of conventional standard-sized satellites, many satellites can achieve precise attitude stabilization (gray group in Figure 1). Nowadays, however, small satellites are also used for sophisticated missions that would previously be performed by standard-sized satellites. Therefore, small satellites also require precise attitude stabilization (orange group in Figure 1). Previously, many studies have focused on precise attitude control for conventional standard-sized satellites. These methods are now being applied to small satellites; however, some of them are unsuitable for standards satellites because of constraints such as power consumption, mass, and space requirements. This study clarifies these constraints and problems and proposes new methods to solve some of these problems with the objective of realizing precise attitude stabilization for nano- and micro-satellites.

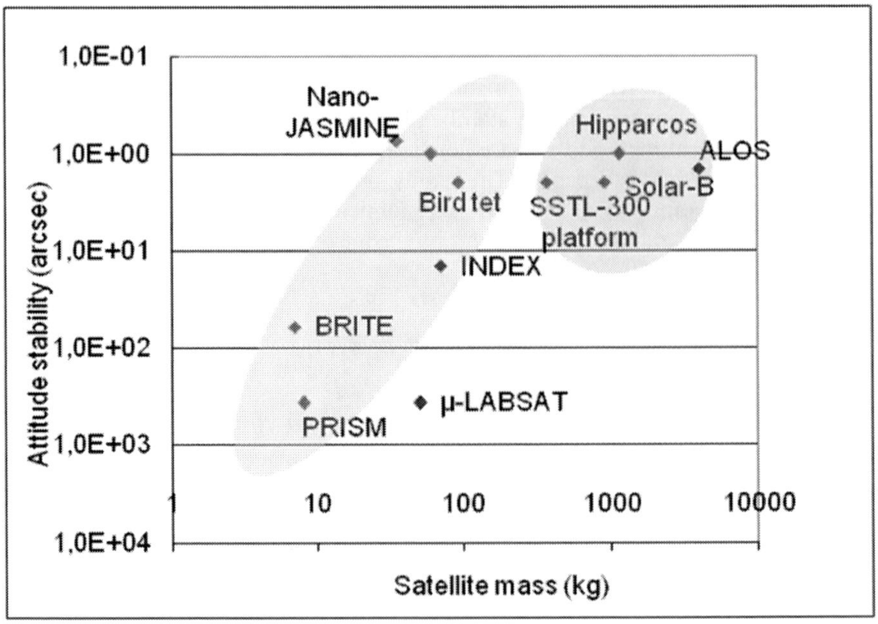

Figure 1. Relationship between satellite mass and attitude stabilization accuracy.

2. Overview of ADCS in Nano-JAMSINE

This study focuses on Nano-JASMINE as an example of small satellites that require precise attitude control. As mentioned above, Nano-JASMINE is an astrometry nano-satellite that was developed by the ISSL in cooperation with the NAOJ. Nano-JASMINE weighs 35 kg and has a volume of approximately 50 cm × 50 cm × 50 cm. Table 1 lists its specifications. Nano-JASMINE will be launched into a sun-synchronous orbit in 2014. It will then perform an all-sky survey during an operational period of approximately two years. Its objective is to measure the three-dimensional positions of stars to an accuracy of three milli-arcseconds (mas) by performing stellar parallax measurements. Nano-JASMINE uses a charge-coupled device (CCD) driven by time delay integration (TDI) to obtain star images. The use of TDI enables Nano-JASMINE to track and observe a large number of faint stars. To capture the star images using TDI, the spin rate of Nano-JASMINE has to be synchronized with the TDI scanning rate to an accuracy of half the size of one CCD pixel (its view angle is approximately 740 mas) during the exposure time (approximately 8.8 s). Thus far, such high-accuracy attitude stabilization has not been accomplished in any small satellite weighing less than 50 kg.

In general, although small satellites afford the above-described advantages, namely, low development cost, short development period, and applicability to various space missions, they are unsuitable for missions with strict requirements because of constraints such as power consumption, mass, and space requirements. In particular, precise ADCS is difficult to achieve in small satellite for the following two reasons. (1) The effect of attitude disturbances is stronger in small satellites than in standard-sized satellites because of their smaller moment of inertia. Magnetic attitude disturbance is the dominant attitude disturbance for low Earth orbit (LEO) small satellites. (2) Some conventional methods are unsuitable for application to small satellites because of constraints such as power consumption, mass, and space requirement. For example, rate estimation using precise gyro sensors such as ring-laser gyroscope (RLG) cannot be applied because of the abovementioned constraints.

Table 1. Specifications of Nano-JASMINE

Item		Value
Satellite	Size	508×508×512 mm^2
	Mass	35 kg
	Orbit	Sun-synchronous orbit
	Mission	Infrared astrometry
Telescope	Focal length	1.67 m
	Diameter	5 cm
	Detector	Time delay integration CCD
Attitude control system	Attitude accuracy	0.01 deg
	Attitude rate accuracy	$4 \times 10^{-7}\ rad/s$ (z-axis) $2 \times 10^{-6}\ rad/s$ (x,y-axis)
	Sensor	Sun sensor, Magnetometer, FOG, STT
	Actuator	RW, MTQ, Magnetic canceller

This study focuses on magnetic disturbance and the constraints faced in small satellites for precise attitude stabilization.

3. Magnetic Attitude Disturbance

3.1. Magnetic Attitude Disturbance

Nano- and micro-satellites in LEO suffer from various attitude disturbances such as magnetic disturbance, air pressure disturbance, solar disturbance, and gravity gradient disturbance. In standard-sized satellites, the effect of magnetic disturbance is insignificant and therefore not considered. However, in small satellites, magnetic disturbance is the dominant attitude disturbance. This is because the magnetic moment has a relatively strong effect on such satellites owing to their small moment of inertia. Furthermore, the geomagnetic field has a large magnitude in LEO (Inamori et al., 2011(d)). To satisfy the strict attitude requirements for nano- and microsatellites in LEO, the effect of magnetic disturbance should be mitigated to achieve precise attitude control. Figure 2 and Figure 3 show the magnetic moment and the effect of magnetic disturbance in previous satellite missions, respectively. In Figure 3, the horizontal and vertical axes show the satellite moment of inertia and the angular acceleration owing to magnetic disturbance, respectively. As shown in Figure 2, although the moment of inertia decreases 1000-fold, the magnetic moment of a satellite decreases only 10-fold. Thus, the magnetic disturbance torque affects smaller satellite attitudes greatly. Figure 3 shows that the angular acceleration caused by a magnetic disturbance torque is larger than that of smaller satellites. Therefore, this effect should be compensated in small satellites. Table 2 presents attitude disturbances and in a small satellite and standard-sized satellites in LEO. The table shows that a magnetic disturbance significantly affects the satellite attitude in smaller satellite missions.

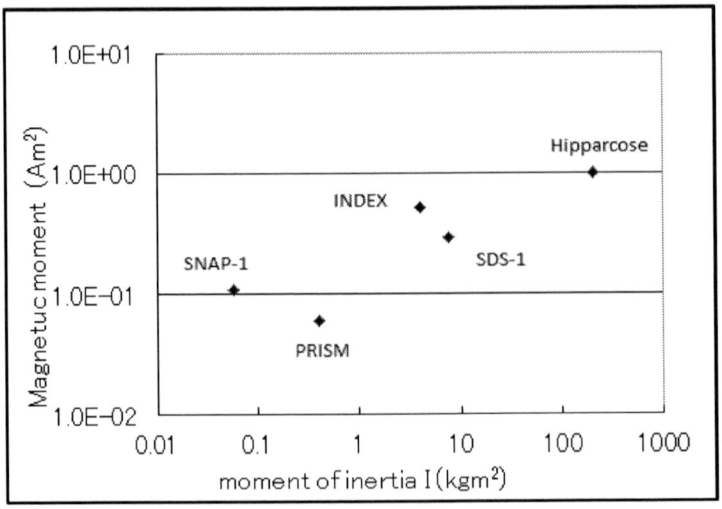

Jorck et al., 1999; Steyn, W.H., and Hashida, Y., 2001; Sakai et al., 2008; Tanaka et al., 2009; Suehiro, 2010.

Figure 2. Relationship between satellite mass and residual magnetic moment of a satellite.

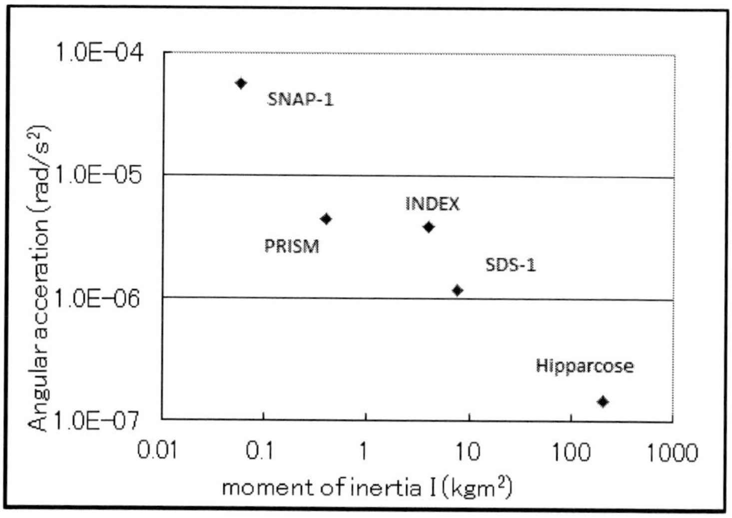

Figure 3. Relationship between satellite mass and residual magnetic moment of a satellite.

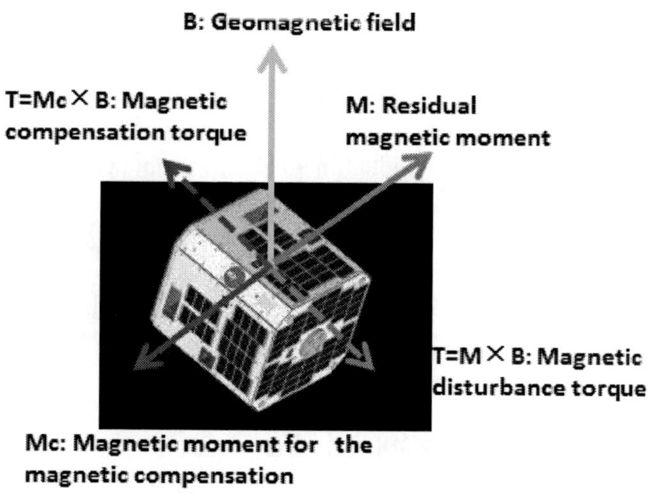

Figure 4. Magnetic moment compensation using MTQs.

Table 2. Attitude disturbances in a small satellite PRISM and a standard-sized satellite ASUKA and ASTRO-E

Disturbance	PRISM (8.5 kg)	ASUKA (420 kg)	ASTRO-E (1567 kg)
Magnetic disturbance	$3 \times 10^{-6}\ Nm$	$5 \times 10^{-5}\ Nm$	$2 \times 10^{-4}\ Nm$
Air disturbance	$3 \times 10^{-8}\ Nm$	$1 \times 10^{-4}\ Nm$	$5 \times 10^{-4}\ Nm$
Solar pressure disturbance	$1 \times 10^{-8}\ Nm$	$4 \times 10^{-5}\ Nm$	$2 \times 10^{-4}\ Nm$
Gravity gradient disturbance	$8 \times 10^{-7}\ Nm$	$1 \times 10^{-3}\ Nm$	$1 \times 10^{-2}\ Nm$

Ninomiya et al., 2003), Inamori et al., 2011(a).

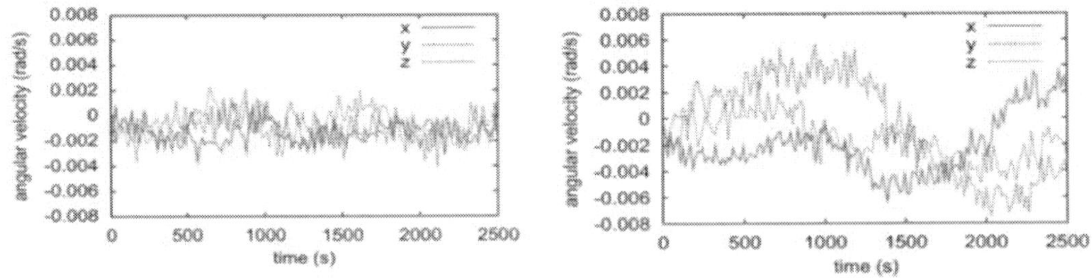

(A) (B)

Inamori et al., 2011(b).

Figure 5. Results of in-orbit experiments for magnetic compensation using remote sensing nano-satellite PRISM: angular rate of PRISM: a) without magnetic compensation and b) with magnetic compensation.

3.2. Magnetic Disturbance Compensation

Magnetic disturbance is caused by the interaction of the geomagnetic field and the residual magnetic moment of a satellite. If a satellite estimates the residual magnetic moment and compensates for it using the steady output of a magnetic torquer (MTQ), it can compensate for the effect of magnetic disturbance. Figure 4 shows an overview of the compensation of magnetic disturbance. To compensate for the magnetic disturbance precisely, a satellite must estimate the residual magnetic moment accurately and must have an accurate actuator to generate a magnetic moment. The satellite can estimate the magnetic moment using an extended Kalman filter with magnetometers and gyro sensors. This method is examined using an in-orbit satellite.

Figure 5 shows the result of magnetic compensation using the remote sensing nano-satellite PRISM. Figure 5-a) shows the attitude rate of PRISM without magnetic compensation: the satellite attitude rate changes relatively rapidly because of the magnetic disturbance effect. With magnetic compensation as shown in Figure 5-b), the satellite can stabilize the attitude precisely because of the relatively small magnetic moment.

The magnetic compensation is effective method for steady magnetic moment, however the method is not effective for no steady high frequency magnetic moment, because the extended Kalman filter is not useful for the estimation of the high frequency magnetic moment. Table 3 shows an overview of the sources of the time-variable magnetic moments in Nano-JASMINE.

Time-variable residual magnetic moments of an orbital frequency are induced by environmental changes such as temperature and the sunlight. For example, sunlight causes current loops in solar batteries, in which the magnetic moment depends on the direction of the sun. Because the change of this frequency is not so rapid, this frequency can be estimated using the extended Kalman filter. Thus the effect of these magnetic moments can be compensated in orbit.

Time-variable residual magnetic moments of a control frequency are induced by an on-board computer and electrical devices in a satellite. In devices connected to the microcomputer, electrical currents change in a calculation cycle of the computer.

Table 3. Sources of the time variable magnetic moments in Nano-JAMSINE

Item		frequency	Magnitude	Method
Heater		Orbital frequency	1×10^{-4}Am^2	In-orbit compensation
Alignment change of magnetic materials			1×10^{-5}Am^2	
Electrical current change caused by temperature change			less than 1×10^{-3}Am^2	
Battery		Orbital frequency +Step change	6×10^{-3}Am^2	
Solar battery			1×10^{-2} - 1×10^{-3}Am^2	
Tx / Rx		Step change	-	
Reaction wheels		Rotation frequency	1×10^{-4}Am^2	Compensation in a design phase
Electrical circuits	for a power subsystem	Control frequency	3×10^{-3}Am^2	
	for sensors		1×10^{-6}Am^2	
	for reaction wheels		1×10^{-5}Am^2	
MTQ for magnetic compensation		All frequencies	2×10^{-6}Am^2	
Electrical wires			less than 1×10^{-5}Am^2	

Thus magnetic moments caused from harnesses generate the time-variable magnetic moments with this frequency. Generally, the magnetic moment of the control frequency is difficult to be estimated using the extended Kalman filter, because the control frequency of the devices is relatively high in most satellite missions. Thus the magnetic moments should be reduced in a design phase on the ground, if the satellite should consider these magnetic moments to achieve attitude requirements.

Small satellites may also have step change magnetic moments. If the satellite moves to other phase, when some devices are turned on, and others are turned off, which cause magnetic moment change. The sunlight also causes step change magnetic moments when the satellite gets into and out from eclipse. These changes are too sudden to estimate using the extended Kalman filter. The satellite needs to estimate the residual magnetic moment in each phase before observation and to switch the value for magnetic compensation in order to reduce this effect.

4. Difficulty Using Conventional Sensors and Actuators

4.1. Constraints for Small Satellites

Because a small satellite has strict constraints of power consumption, mass, and space requirements, some conventional methods for precise attitude control, such as rate sensing and jitter reduction of a reaction wheel (RW), are not applicable to them.

In these satellites, RWs are used as actuators for precise attitude control. Because RWs have an easily measurable speed that provides very precise control of their rotation speed, they permit very precise changes in a spacecraft's attitude. These actuators enhance the accuracy of attitude control by reducing the effect of low-frequency disturbances; however, the RWs degrade the attitude control accuracy at high frequency because of the jitter generated by the wheel imbalance and bearing irregularities. Such jitter degrades the quality of scientific data or Earth images; therefore, jitter reduction is essential for achieving sophisticated missions such as astronomical observations or remote sensing. In previous large satellite missions, tip-tilt mirrors (TTMs) and isolators were used for controlling or isolating the vibration; however, these devices are difficult to use in small satellite because of the strict constraints of power consumption, space requirements, and mass.

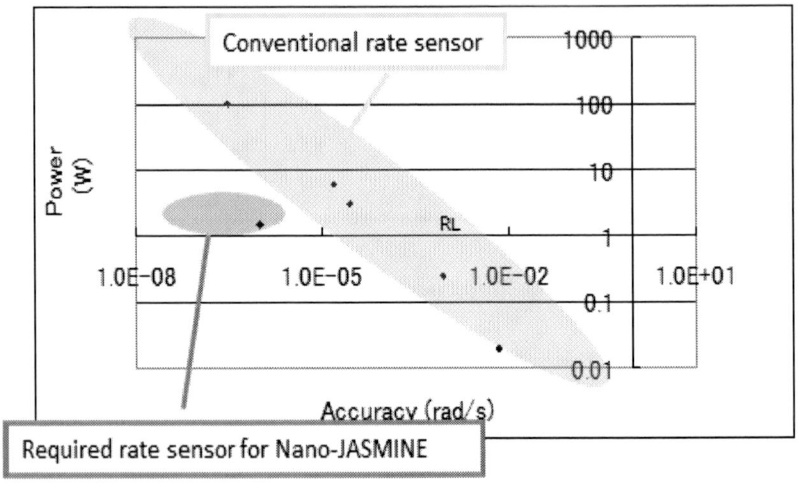

(A)

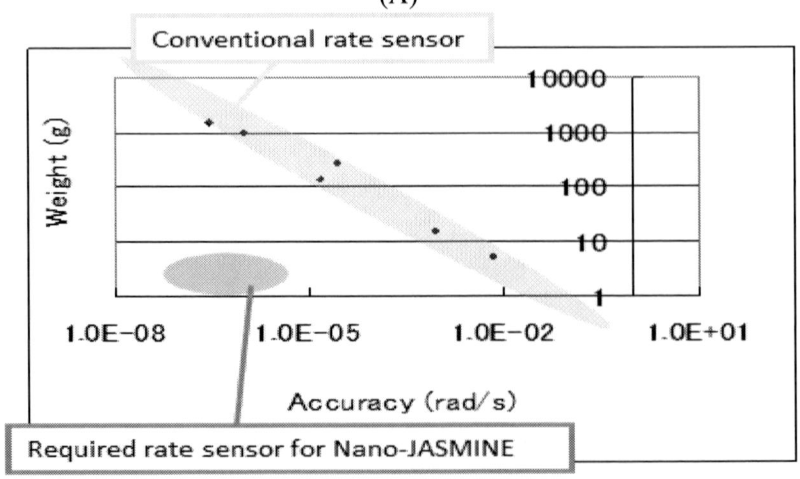

(B)

Figure 6. Power consumption and weight of a rate sensor: a) power consumption vs. accuracy of a rate sensor and b) weight vs. accuracy of a rate sensor.

Further, in these satellites, conventional FOG and RLG are also difficult to use because of these constraints. Figure 6 shows the relationship between the accuracy and the power consumption as well as mass of a rate sensor. The figure shows that an accurate rate sensor for precise attitude control has a large power consumption and mass. Thus, in general, accurate rate sensors are not available for the precise attitude control of a small satellite. For example, Nano-JAMSINE requires a stabilization of 1×10^{-6} rad/s during missions with a 30-W power system, which is difficult to achieve using FOG and RLG, as shown in Figure 6. Therefore, a new method to determine the angular rate accurately is indispensable for small satellites.

4.2. Jitter Reduction Using Small Reaction Wheels

In many satellites, the jitter of an RW degrades the accuracy of attitude control. To reduce the effect of jitter, TTMs and isolators are used in many satellites (Agrawal et al., 2004;Davis et al., 1986; Davis et al., 1994). For example, Solar-B, a solar observation satellite developed by the Japan Aerospace Exploration Agency (JAXA), uses a TTM for observations (Kosugi et al., 2007). The objective of the Solar-B mission is to explore the magnetic field of the Sun, which requires attitude stabilization to an accuracy of 60 mas per 10 s. To achieve this accuracy, the effect of disturbances due to RWs should be reduced. To suppress high-frequency jitter, a TTM was mounted on a solar optical telescope. TTMs mirrors are effectively segmented mirrors having only one segment that can tip and tilt, rather than an array of multiple segments that can tip and tilt independently. By controlling a part of the satellite body using the TTM as opposed to flexibly controlling the entire satellite body, the telescope stabilization requirement can be achieved rather easily.

Both TTM and isolators are effective in reducing the effect of the jitter from reaction wheels; however, they are difficult to use for small satellites because of the power consumption, space requirement, and mass constraints.

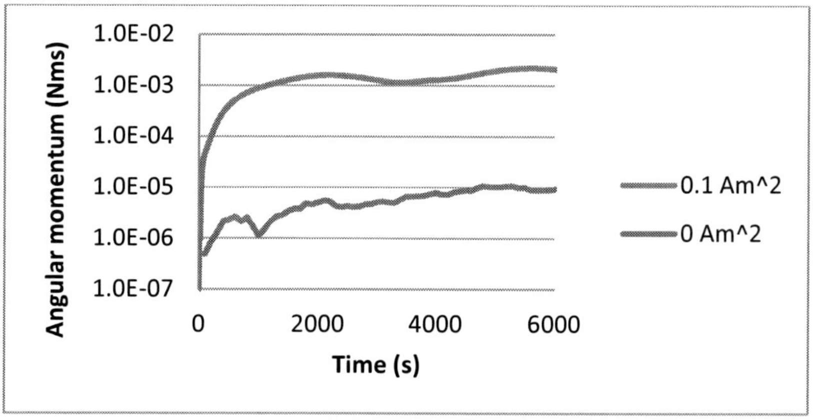

Figure 7. Angular momentum of reaction wheels: with magnetic compensation and without magnetic compensation.

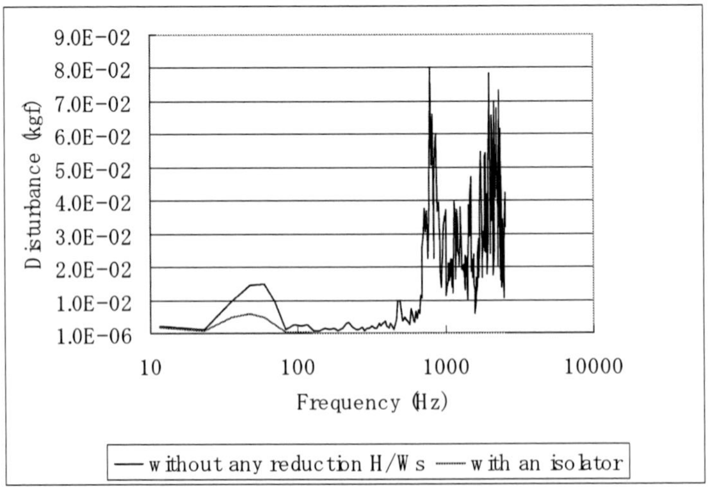

(A)

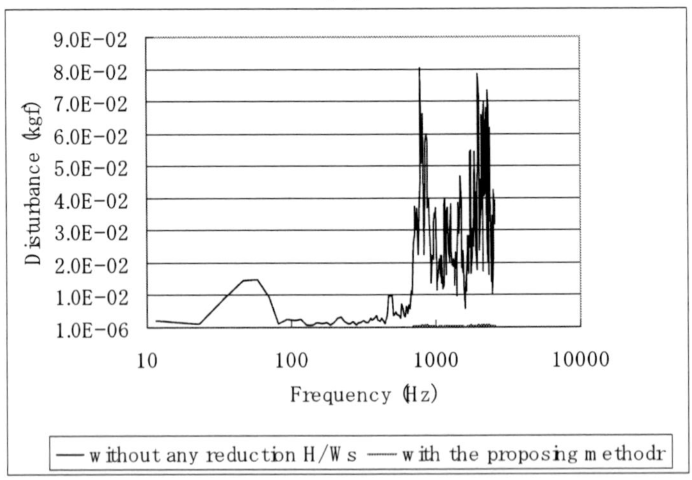

(B)

Figure 8. Jitter reduction of a RW: a) with an isolator and b) with the proposed method.

In this study, the jitter of RWs is reduced by using small RWs, and concurrently, the magnetic disturbance is compensated using MTQs. With magnetic compensation, the satellite reduces the angular momentum that RWs obtain; therefore, the satellite does not need large RWs that may otherwise cause jitter. Figure 7 shows the simulation result of the RW angular momentum in orbit. In this simulation, a satellite stabilizes the attitude using RWs; therefore, the RW obtains angular momentum from attitude disturbances such as the magnetic disturbance. As the blue line in Figure 7, without magnetic compensation, the RW obtains angular momentum from disturbances. In order to store the angular momentum in the RW, the satellite must use relatively larger RWs, which usually cause larger jitter. The red line in Figure 7 shows the angular momentum of an RW with magnetic compensation. Because the satellite reduces compensates for the magnetic disturbance using MTQs, the RW obtains a small magnitude of angular momentum. Therefore, the satellite can use a relatively small RW, which does not cause jitter. As a result, the satellite can reduce the effect of jitter without the use of conventional isolators and TTMs. Thus, by using this method, the satellite can achieve precise attitude control under the constraints of low power consumption, small space, and low mass. Figure 8 shows the result of jitter reduction by using a conventional method using an isolator (a) and using the proposed method (b). In Figure 8, in the case of the conventional method, the jitter of an RW with an isolator is calculated using experimental data and the expected transform function of the isolator. As shown in Figure 8, the isolator is effective in a high-frequency area; however, it is not effective in a low-frequency area. Therefore, the isolator cannot reduce relatively low-frequency jitter produced by a certain rotation frequency. Low-frequency jitter interacts with the detector used for observations and degrades the scientific data. By using the proposed method, jitter reduction is effective at all frequencies, and therefore, the scientific data is not degraded by the jitter. In addition, by using this method, the total mass and power consumption of the satellite can be kept low through the use of small RWs; therefore, the method is useful for small satellites that have strict constraints.

4.3. Rate Estimation Using Astronomical Bodies

Small satellites cannot use precise attitude rate sensors, as shown in Figure 6. This research proposes a method for estimating the angular rate by using star images obtained from an on-board mission telescope. In this case, the satellites do not need install additional rate sensor for precise attitude rate control. This method uses the blurriness of star images. If the satellite attitude is unstable, the obtained star images are blurry and not sharp. Figure 9 shows how a satellite estimates the angular velocity using a star image. First, the satellite calculates the line spread function (LSF), which is a star image compressed to two axes (horizontal and vertical axes). The blurriness of a star can be assessed using the variance of the LSF, which changes with the satellite attitude rate. Figure 10 shows the relationship between the angular rate of the satellite and the star images. If the satellite attitude has an angular velocity, the star images obtained from the telescope are blurred in the direction of the axis of rotation.

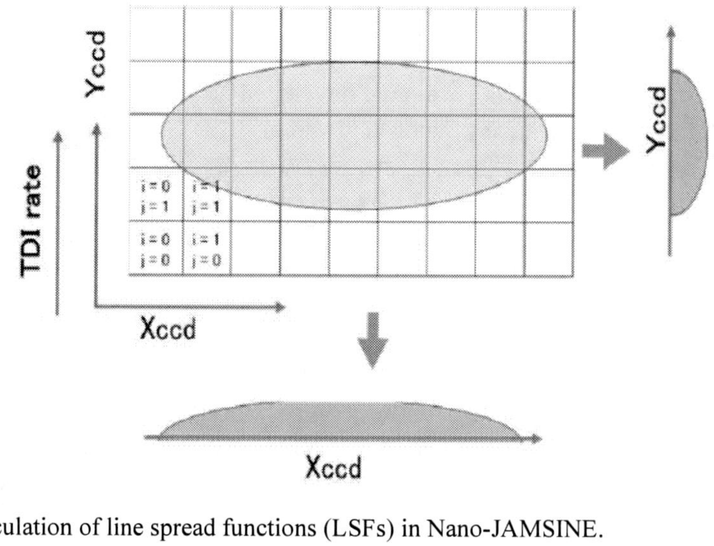

Figure 9. Calculation of line spread functions (LSFs) in Nano-JAMSINE.

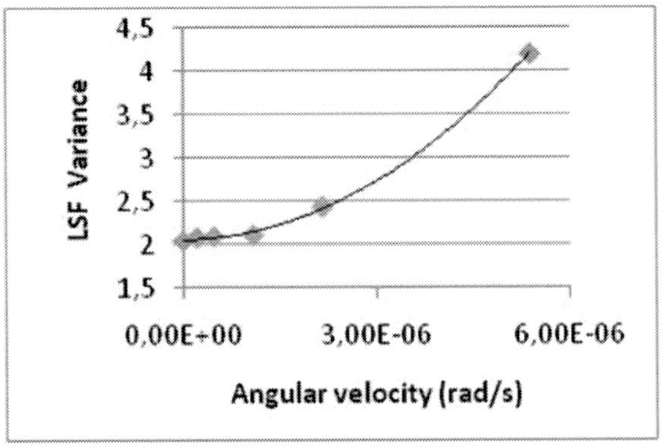

Figure 10. Relationship between angular velocity and LSF variance.

From these LSFs, variances of a star image can be obtained as follows. In this calculation, a star image is assumed to be extracted from FOV by 5 9 pixels.

(1)

In the same manner, the LSF variance for Y axis can be calculated as follows,

(2)

Where i and j are pixel IDs, f(I, j) is the magnitude of a star image at pixel ID i and j, respectively. The relationship of the angular rate of a satellite and the variance of a LSF can be depicted as follows,

(3)

is a function of the exposure time of a CCD, is a constant which shows the LSF variance when the satellite is ideally stabilized, and is the angular rate of the satellite in the TDI frame, respectively.

For the verification of the method, we have developed a test system. In this test system, star images are imitated with a collimator and small LED. The LED lighting time is controlled strictly under constant TDI rate. In this situation, the satellite obtains bluer star images caused by the TDI velocity. With the blurriness of a star image, the TDI rate can be estimated. Figure 10 shows the relationship between LSF variance and angular velocity obtained by this experiment. The figure shows that the relationship between blurriness of a star image and angular velocity can be expressed as a quadratic function which is shown in Equation 3.

Figure 11 and 12 show simulation results of angular rate estimation and attitude stabilization. In these simulations, the satellite angular velocity is estimated using an EKF with STT and FOG measurements in the initial phase before 4000 sec. After the 4000 sec, the angular rate is estimated using the proposed method. These figures show that the angular rate of the satellite after 4000 sec is more accurately estimated and stabilized than those of before 4000 sec.

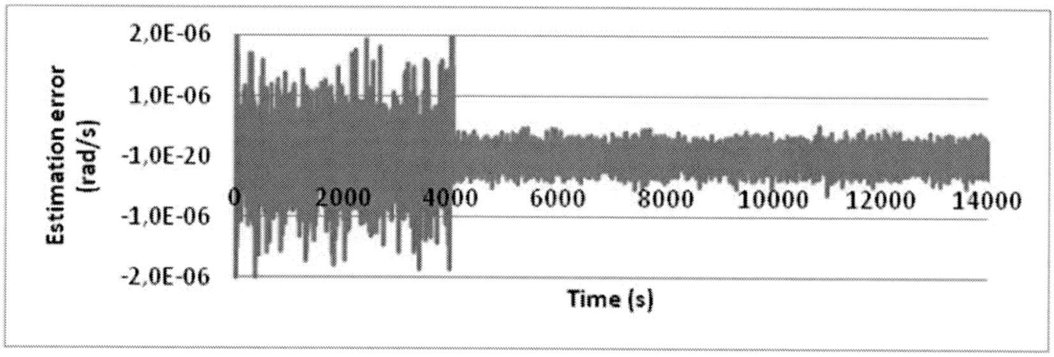

Figure 11. Simulation result of angular rate estimation.

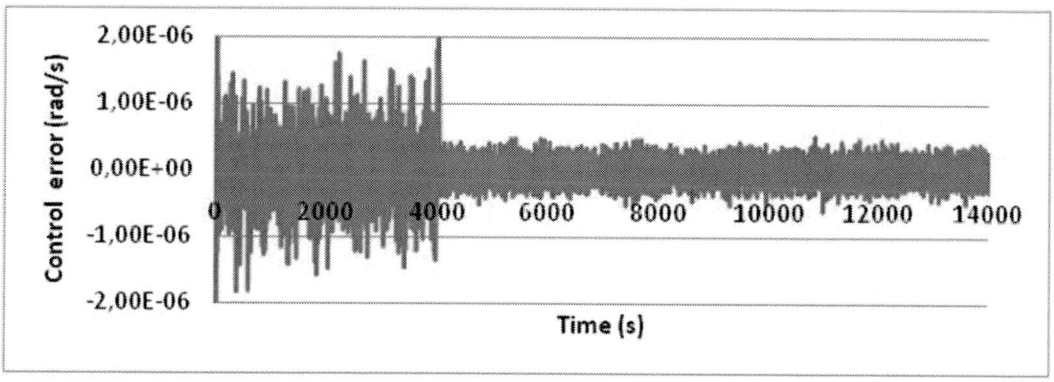

Figure 12. Simulation result of angular rate control.

From Figure 11, the angular rate can be estimated to an accuracy of rad/s with the root-mean-square. From Figure 12, the angular rate can be controlled to an accuracy of rad/s, which meets the requirements of attitude stabilization to an accuracy of 740 mas during 8.8 s exposure time. With the proposed method, the satellite can estimate and control the angular velocity without any additional rate sensors.

CONCLUSION

In this study, we focused on the development of a precise attitude stabilization system for small satellites. For the precise attitude stabilization, several problems had to be considered. The first problem was the effect of magnetic disturbance. Because of the small moment of inertia of a small satellite, the magnetic disturbance degraded the accuracy of attitude control. In this study, the magnetic disturbance was compensated by using the steady output from MTQs. The second problem was the constraints of power consumption, space requirement, and mass. Because small satellites are generally designed at low cost within a short period, they involve strict constraints. Therefore, some conventional methods that are used in standard-sized satellites are not applicable to small satellites. In this study, we focused on the jitter of an RW and rate estimation. To reduce the jitter of the RW, we used a small RW in conjunction with magnetic compensation. To develop a precise attitude rate sensor, we used the blurriness of star images obtained from an on-board telescope. Simulations of these methods showed that precise attitude control was feasible in small satellites with the proposed methods.

ACKNOWLEDGMENTS

The author would like to thank the members of the Nano-JASMINE project at the University of Tokyo for their unmatched technical competence and assistance.

REFERENCES

Agrawal , Brij N., and Chen , H.,:"Algorithms for active vibration isolation on spacecraft using a Stewart platform", Smart Material and Structures, Vol. 13, pp. 873-880, 2004.

Davis , L. P., Wilson, J. F., Jewell , R. E. : "Hubble Space Telescope Reaction Wheel Assembly Vibration Isolation System," Honeywell Inc., 1986

Davis, P., Cunningham, D., Harrell, J. : "Advanced 1.5Hz Passive viscous isolation system", AIAA/ASME/ASCE/AHS/ASC Structures, Structural Dynamics and materials Conference 35th, 1994

Heidt, H., Puig-Suari, J., Moore, A. S., Nakasuka, S., Twiggs, R. J., CubeSat: a new generation of pico satellite for education and industry low-cost space experimentation, 14th Annual AIAA/USU Conference on Small Satellites,Logan,2000.

Inamori, T., Sako, N., & Nakasuka, S. (2011). Magnetic dipole moment estimation and compensation for an accurate attitude control in nano-satellite missions, *Acta Astronautica*, Volume 68, Issues 11-12, (2011), pp2038-2046.

Inamori, T., Shimizu, K., Mikawa, Y., Tanaka, T., & Nakasuka, S. (2011). Attitude stabilization for the nano remote sensing satellite PRISM, *ASCE's Journal of Aerospace Engineering*, Article in press, doi: 10.1061/(ASCE)AS.1943-5525.0000170.

Inamori, T., Sako, N., & Nakasuka, S. (2011). Attitude control system for the nano-astrometry mission "Nano-JASMINE"", *Aircraft Engineering and Aerospace Technology*, (2011),Volume 83, Issues 4, pp.221 – 228.

Inamori, T., Sako, N., & Nakasuka, S. (2011). Compensation of time-variable magnetic moments for a precise attitude control in nano- and micro-satellite missions, *Advances in Space Research*, (2011), Volume 48, Issue 3, pp 432-440.

Jorck H., HolotaW., Hupfer W., Viehmann D, Control System Approaches for an Advanced Astrometry Satellite, Proceedings of the 4th ESA International Conference on Spacecraft Guidance Navigation and Control Systems, 18 October - 21 October, Noordwijk, Netherlands, 1999.

Kosugi , T., Matsuzaki, K., Sakao , T., Shimizu, T., ·Sone, Y., Tachikawa, S., Hashimoto,·T., K. Minesugi, A., Ohnishi, Yamada, T., Tsuneta , S., Hara , H., Ichimoto , K., Suematsu, Y., Shimojo, M., Watanabe, T., Shimada, S., Davis, J.M., Hill,·L.D., Owens, J.K., Title, A.M., Culhane , J.L., Harra, L.K., Doschek , G.A., Golub, L., The Hinode (Solar-B) Mission: An Overview, SOLAR PHYSICS, Volume 243, Number 1 (2007), 3-17, DOI: 10.1007/s11207-007-9014-6, 2007.

Nagai, M., Eishima, T., Nakamura, Y., Funase, R., Enokuchi, A., Funane, T., Nojiri, Y. , Sasaki, F., Nakasuka, S., University of Tokyo's PICO-SATELLITE PROJECT ''PRISM,'' 56th International Astronautical Congress, Fukuoka, 2005.

Nakamura, Y., Eishima, T., Nagai, M., Funase, R., Enokuchi, A., Nakada, K., Cheng, Y. , Takei, E. T., Funane, T., Sasaki, F., Nojiri, Y., Yamamoto, T., Nagayama, E. , Nakasuka, S., University of Tokyo's Ongoing Student-Lead Pico-Satellite Projects—CubeSat XI and PRISM, 55th International Astronautical Congress, British Columbia, 2004.

Ninomiya, K., Hashimoto, T., Maeda, K., Saitoh, T., Akisue, M., Genba, A., "Attitude Control Systems of X-Ray Astronomy-Mission Satellites ""ASCA(ASTRO-D)"" and ""ASTRO-E""", *Institute of Space and Astronautical Science report*, 121, 1-213, 2003-03.

Sakai, S.-i. Fukushima, Y. Saito, H. (2008). Design and on-orbit evaluation of magnetic attitude control system for the "REIMEI" microsatellite, Advanced Motion Control, 2008. AMC '08. *10th IEEE International Workshop* on , pp.584-589, 26-28 March 2008.

Steyn, W.H., and Hashida, Y., In-Orbit Attitude Performance of the 3-Axis Stabilised SNAP-1 Nanosatellite, paper presented at 15th Annual AIAA/USU Conference on Small Satellites, 13 August-16 August, Logan, Utah, 2001.

Suehiro, T., Satellite Design Methodology to Suppress Time-varying Residual Magnet Effects on Attitudefor Nano-Satellites, 18th *IFAC Symposium on Automatic Control in Aerospace*, 6 September-10 September, Nara Japan, 2010.

Tanaka T., Inamori, I., Sako, N., Nakasuka, S., Magnetism Management for the Nano-JASMINE mission, *The 27th International Symposium on Space Technology and Science*, 26 July-17 July, Tsukuba Japan, 2009.

In: Recent Advances in Satellite Research and Development ISBN: 978-1-62417-443-8
Editors: Samuel Gardiner and Kenneth P. Olsen

Chapter 7

THIRD GENERATION GPS: A LOW-MAINTENANCE, HIGH-RELIABILITY FUTURE GPS SYSTEM

Ezzat G. Bakhoum
Department of Electrical and Computer Engineering
University of West Florida, Pensacola, FL, US

Abstract

The third-generation Global Positioning Satellite system (GPS system) is now in the early development phase by Celestia Corporation. The new system, major components of which were designed by the author, has substantial advantages by comparison with the current GPS system. Doppler correction and signal tracking circuitry in the receivers have been eliminated, which resulted in a substantial reduction in the size of the receiver (in the near future, a GPS receiver will fit inside an ordinary wrist watch). Maintenance signals, which are frequently transmitted from the earth stations to the satellites, have also been substantially reduced. Simulations that characterize the User Equivalent Range Errors show the advantages of the new system. Overall, the new system is substantially more reliable, offers better performance, and is substantially cheaper than the current GPS system. The appendix to this chapter gives a closed-form solution for the GPS geolocation equations.

1. Overview and Introduction

The author was a member of the team that developed the second-generation GPS system at Lockheed Martin / L3 Communications during the late 1990s. That system of satellites (officially known as the Block IIR system) has recently been deployed [1]. The objective of this chapter is twofold: first, to introduce a new, third-generation GPS system that is now in development by Celestia Corp. (a privately held corporation, with approximately $1.5 Billion in funding). The second objective of this chapter is to present a closed-form solution for the GPS geolocation equations that does not depend on range data (a mathematical development that will allow the GPS receiver to be miniaturized without sacrificing accuracy). Although the second-generation GPS system developed by Lockheed Martin is quite sophisticated, the planned new system will further significantly advance the accuracy and the reliability of GPS technology. Specifically,

- The third-generation system will eliminate the need for Doppler correction circuitry as well as signal tracking and synchronization circuitry in the receivers. This radical improvement, together with the closed-form solution of the geolocation equations presented in this chapter, will allow the size of the GPS receiver to be reduced to the size of a typical wrist watch.
- The new system will allow the satellites to operate in relativistic-null orbits; which will substantially reduce the amount of maintenance and timing correction signals that must be transmitted from the control stations to the satellites.
- Due to the above improvements, the new system will be able to use high bit-rate QPSK or Offset QPSK signals, versus the low bit-rate (50 bits per sec.) BPSK signal that is currently used in the second-generation system.

Figure 1 shows the fundamental principle of the new system.

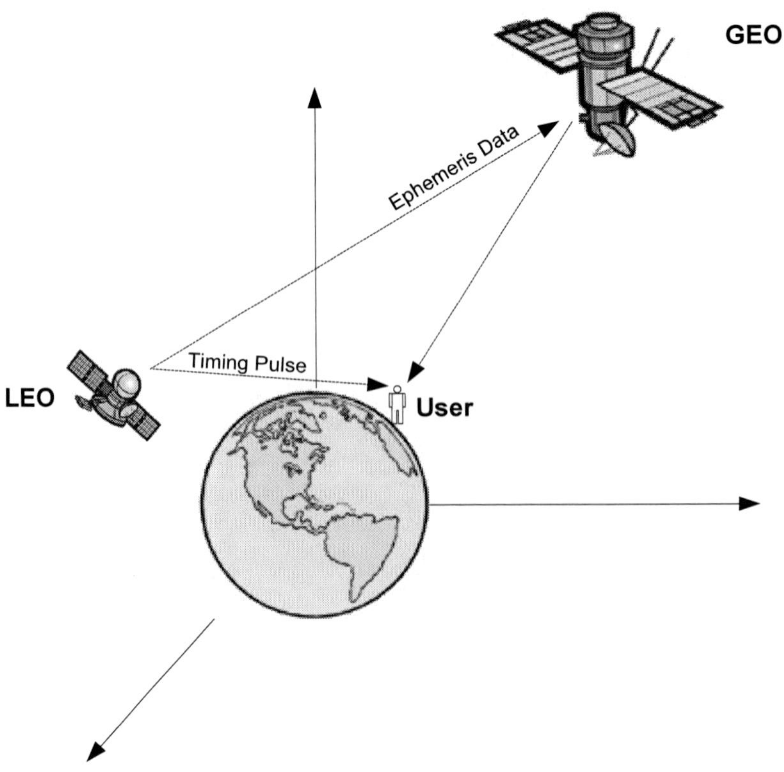

Figure 1. The new planned third-generation GPS system. A group of 36 LEO satellites in relativistic-null orbits transmit timing pulses to the receivers via dedicated analog channels. The digital ephemeris and almanac data is subsequently relayed to the receivers via a Geostationary satellite. The system is shown in an earth-centric, earth-fixed coordinate system.

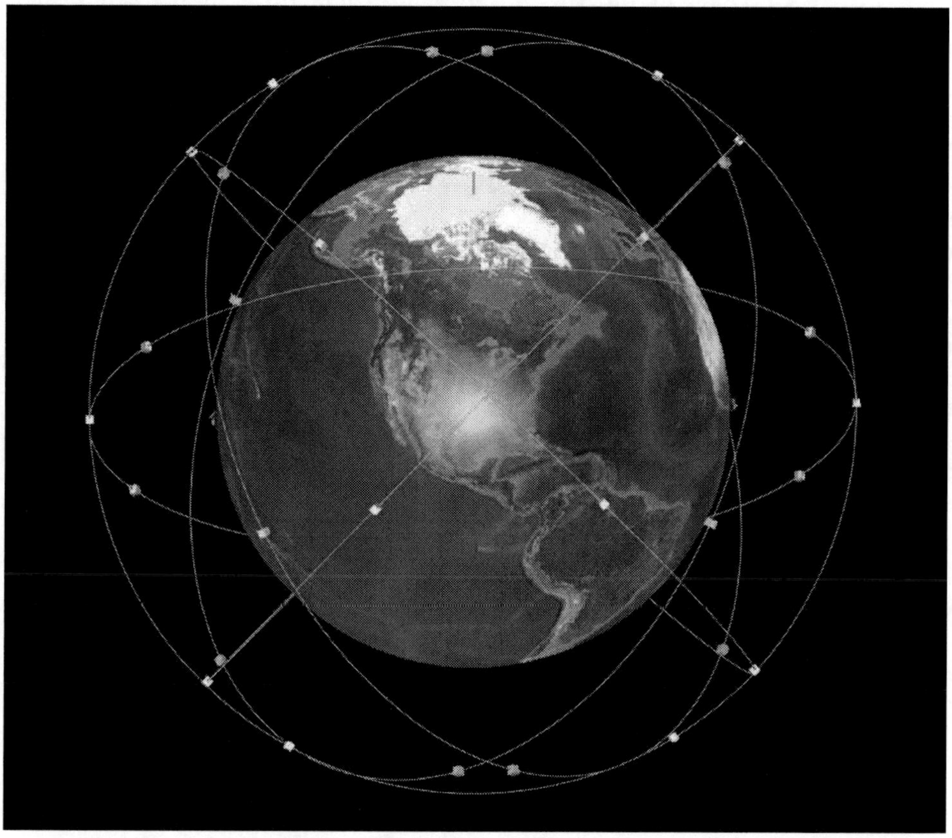

Figure 2. The constellation of 36 LEO satellites.

As shown in the figure, a group of 36 LEO satellites (here, only one is shown for clarity) will be placed in inclined, relativistic-null orbits. A relativistic-null orbit is a medium earth orbit that has an altitude of 3180 km (we will refer to the satellites as LEOs, however, for simplicity). In a relativistic-null orbit, the two relativistic effects of time dilation and gravitational blueshift cancel each other [2, 3]. This is a very desirable feature, since most of the maintenance signals that are transmitted from the control stations to the GPS satellites are used to correct the timings of the satellites' on-board atomic clocks, which constantly drift due to the relativistic effects. Hence, by placing the GPS satellites in relativistic-null orbits, the need for constantly monitoring and adjusting the timings of the clocks in those satellites will be reduced greatly. The downside of using medium earth orbit satellites, on the other hand, is that the beam coverage area is smaller by comparison with high-altitude satellites, and hence more satellites are usually needed. The system is designed such that a minimum of 5 satellites will be in view to the receiver at any given time (see the analysis in Section 2). Simulation software has shown that this requirement can be satisfied with 36 satellites placed in 6 orbital planes (the inclination of the orbital planes is not a critical factor, but is presently chosen to be 55°). The simulated constellation is shown in Fig. 2.

In the current GPS system, other major problems (in addition to the maintenance signals) include the Doppler shift and the necessity to track and lock on the digital spread-spectrum signals of the satellites. Because the receiver's circuitry must process the signals under such conditions, the receivers are large and the delay for obtaining a first fix is usually quite significant. These problems are solved in the new system by transmitting to the receivers simple timing pulses through 36 dedicated analog channels. The timing pulses will be used only to obtain the time-of-arrival information for the signal of each satellite (and hence the 36 analog channels that the receiver will incorporate will not have any processing capability beyond the simple determination of the time of arrival of each pulse). In order to solve the geolocation equations, however, the ephemeris data (from which the coordinates of the satellites are calculated) is of course also needed. The important observation, however, is that the ephemeris data is not time critical and can be received after a small delay. Accordingly, the LEO satellites will transmit the digital ephemeris (and other data, such as the almanac) to a Geostationary (GEO) satellite, which will then relay the digital data to the users on the ground via one dedicated digital channel. The advantages of using a GEO as a relaying mechanism for the digital data are the following:

- Because the GEO appears to be stationary, the user's equipment will not have to incorporate Doppler correction circuitry[1].
- The GEO, by broadcasting on a single dedicated digital channel, will further eliminate the need for the user's equipment to track and lock on the signals from several different satellites, which is the procedure used in the current system. This will result in further substantial reduction in the circuitry of the receiver.
- Due to the above improvements, a substantially higher bit-rate and an improved modulation scheme (that is, QPSK versus BPSK) will be possible. This will allow the receivers to operate in near "real time", which can be critical in some applications.

A Geostationary satellite covers slightly less than one-half of the surface area of the globe [4], so a minimum of three GEOs will be needed for universal coverage. Still, two remote areas in Antarctica and the North Pole will not receive any coverage unless optional Geosynchronous satellites are added to the system. It is yet to be determined whether that option will be implemented in the new system. A note about the relativistic effects and their impact on GPS performance is now in order. To maintain the geolocation accuracy that is available today, a GPS atomic clock must maintain a fractional time stability of better than one part in 10^{13} [2]. The relativistic effects are approximately 10,000 times as large, and can easily render the system useless if careful measures are not implemented. The relativistic time dilation causes the frequency of a moving clock to be slow by a factor of [2, 3]

$$\frac{\Delta f}{f} = \frac{v^2}{2c^2}, \tag{1}$$

where v is the satellite's velocity in orbit. This is a net "redshift" in frequency. In addition, a gravitational "blueshift" occurs when a clock is simply placed at a high altitude

[1]Note that for receiving the simple timing pulses from the LEOs, the analog channels in the receiver only need to have suitable bandwidth.

in earth's gravity field. This blueshift, which causes the clock to run faster, is given by [2, 3]

$$\frac{\Delta f}{f} = \frac{\Delta U}{c^2}, \tag{2}$$

where ΔU is the Newtonian gravitational potential difference between the satellite and the earth's geoid. The gravitational blueshift is the dominant effect in the current GPS system (altitude = 20,000 km). As was pointed out, the two relativistic effects precisely cancel each other at an altitude of 3180 km. This is obviously a great advantage in the new system.

2. Simulations and Performance Estimation

In an error-free world, the set of geolocation equations derived in the Appendix can be used to compute the receiver's coordinates precisely as was pointed out. In practice, however, a number of sources of error contribute to the so-called User Equivalent Range Error (UERE). Those sources of error are: Ionospheric and Tropospheric effects; multipath distortion effects; satellite clock errors; ephemeris errors; and signal arrival errors. Only the last three sources of errors will be substantially different in the new system. We shall now examine in detail the errors associated with the satellites' clock drift, with the ephemeris data, and with signal arrival.

2.1. Satellite clock drift

As was pointed out in Sec. 1, the two relativistic effects of time dilation and gravitational blueshift will precisely cancel each other in the new system. However, due to earth's nonuniform mass distribution and its multipole moments, the gravitational potential U is in reality not constant at any one orbit. Very small variations in U along the orbit of the satellite are expected; and hence, satellite clock errors, albeit small, are still expected. To estimate such errors, we must calculate the variations in the quantity U/c^2 along the orbit of the satellite (see Eq.2). We first point out that U is given by [5, 6, 7]

$$\begin{aligned} U(r) &= \frac{GM}{r} \sum_{l=1}^{\infty} \sum_{m=1}^{\infty} \left(\frac{a}{r}\right)^l (C_{lm} \cos m\lambda + S_{lm} \sin m\lambda)\, P_{lm}(\sin\phi) \\ &\quad - \frac{GM}{r} \sum_{l=1}^{\infty} \left(\frac{a}{r}\right)^l J_l P_{l0}(\sin\phi) \end{aligned} \tag{3}$$

where G is the gravitational constant, M is earth's mass, r is the radius of the orbit, a is earth's mean radius, C_{lm} and S_{lm} are Stoke's coefficients (or the normalized multipole moments of earth's gravity field), P_{lm} are the normalized associated Legendre functions of degree l and order m, λ is the longitude angle, ϕ is the geocentric latitude angle, and where J_l are the zonal harmonic coefficients associated with earth's ellipsoidal shape. The coefficients J_l, C_{lm} and S_{lm} are well known and extensively tabulated in the literature [5, 6]. Accordingly, the variations in U can be calculated along any given orbit with sufficient precision. On that basis, a Monte Carlo simulation was performed in MATLAB to compute the drift in the orbital clock time (which is proportional to $\Delta U/c^2$), and accordingly calculate the expected UERE from equations (15), (16), and (17). The result of that simulation is shown in Fig. 3.

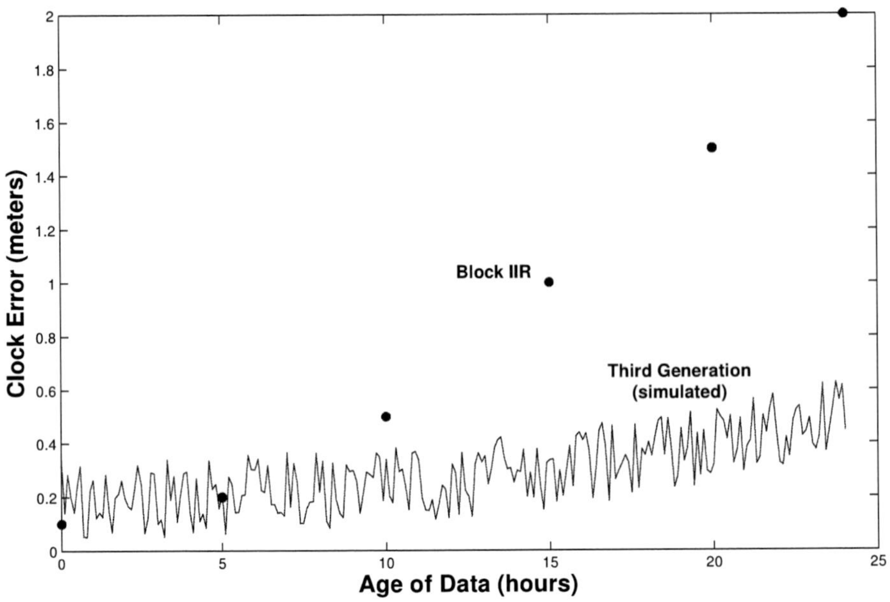

Figure 3. The UERE, in meters, associated with the satellites' clock drift in the Block IIR GPS system (actual), and in the third generation GPS system (simulated).

As shown in Fig. 3, the UERE for the current Block IIR system[2] and for the future third-generation system are shown as a function of the age of data (between ground station uploads). The results of the Monte Carlo simulation for the new system show the average UERE for the constellation of 36 satellites. As the simulation shows, the maximum expected UERE over a 24 hr period is approximately 0.5 m; versus 2 m for the current system. Clearly, frequent data uploads will not be necessary in the new system.
It is to be pointed out that high precision atomic clocks used in GPS satellites do have a natural drift rate; however, such a rate is usually negligible by comparison with the drift associated with the relativistic-geodetic effects discussed here [2].

2.2. Ephemeris errors

The ephemeris data is a set of real-time parameters that describe the orbit of the satellite. The receiver computes the coordinates of the satellite in the earth-centered, earth-fixed (ECEF) frame by using those parameters. The parameters of importance for computing the satellite's coordinates include the mean anomaly M, the eccentricity of the orbit e, the semi-major axis of the orbit a_s, the argument of the perigee ω, the inclination angle i, and a set of 6 coefficients that constitute harmonic correction terms for i, ω, and the radius of

[2]The UERE for the newer, Block IIF (or follow-on) satellites is only slightly different from the data shown for the Block IIR system [11].

the satellite's orbit r. From the ephemeris data, the radius of the orbit r as well as a set of secondary parameters such as the true anomaly ν and the right-ascension angle Ω are calculated. This procedure is described elsewhere [8]. Once the secondary parameters are calculated, the coordinates of the satellite in the ECEF frame are calculated from

$$\begin{aligned} X_{sat} &= r\cos\Omega\cos(\nu+\omega) - r\sin\Omega\cos i\sin(\nu+\omega) \\ Y_{sat} &= r\sin\Omega\cos(\nu+\omega) + r\cos\Omega\cos i\sin(\nu+\omega) \\ Z_{sat} &= r\sin i\sin(\nu+\omega) \end{aligned} \tag{4}$$

Typically, the magnitude of the errors in the ephemeris data is about one part in 10^7 [9, 10]. In the current GPS system (where the satellites have an altitude of 20,000 km) the errors in the ephemeris data result in a UERE of about $\pm$ 2.5 m [9, 11]. For the new system, however, the altitude of the satellites is much smaller; specifically, 3180 km. As a rough estimate, an error of one part in 10^7 based on this altitude should result in a UERE of about $\pm$ 0.3 m. To obtain a more rigorous estimate, however, another Monte Carlo simulation was performed in Matlab. Given the expected geometry of the satellite's orbit, random errors were injected into M, e, a_s, ω, and i; and the errors in the satellite's coordinates were computed from Eqs.(4) above. The UERE was subsequently calculated from the geolocation equations developed in the previous section. The result of that simulation is shown in Fig. 4.

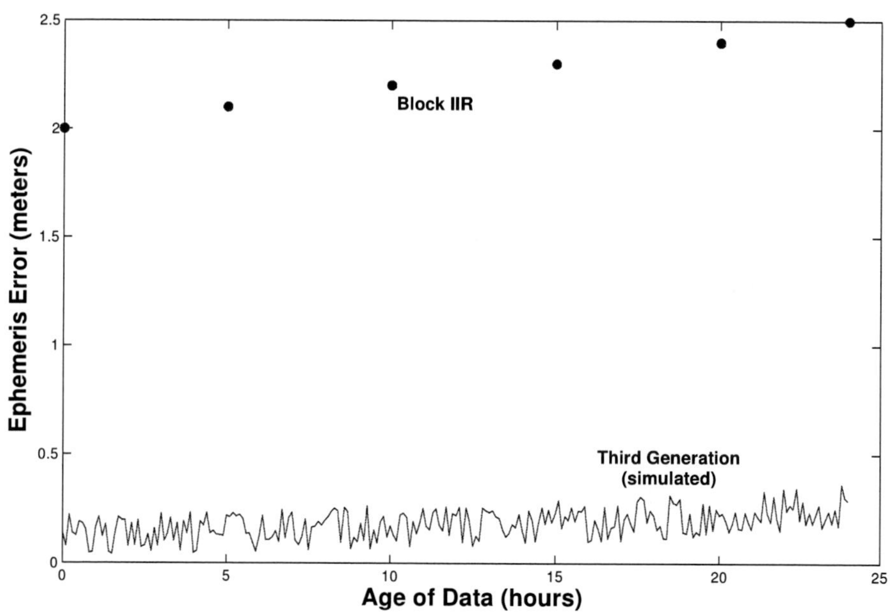

Figure 4. The UERE, in meters, associated with the ephemeris errors in the Block IIR GPS system (actual), and in the third generation GPS system (simulated).

As Fig. 4 shows, the expected UERE is in fact about $\pm$ 0.3 m; significantly smaller than the corresponding error in the current system.

2.3. Signal arrival errors

In the current GPS system, the receiver must recover the spread-spectrum signals of the different satellites and obtain the time-difference-of-arrival between each pair of those signals via digital sequence (pattern) matching. This process (which is necessary to eliminate the bias in the receiver's own clock) has an inherent resolution of $\pm$ 10 ns [8, 12]. Given that the speed of light c is 3×10^8 m/s, the signal arrival error is thus $\pm$ 3 m in the current system. In the new system, however, the timing signals are simple analog pulses[3], as indicated in Sec. 1. It is well known that the resolution inherent in the detection of the leading edge of an analog pulse is about 0.1 ns [13]. This translates to an expected signal arrival error in the new system of $\pm$ 0.03 m; essentially negligible.

2.4. Comparison of the User Equivalent Range Errors in the second and the third generation GPS systems

Table 2.4. shows a comparison of all the sources of UERE in the existing and in the planned GPS systems.

Table 1. Sources of User Equivalent Range Errors in the second and in the third generation GPS systems

Source of UERE	2nd Generation GPS	3rd Generation GPS
Ionospheric Effects	± 5 m	± 5 m
Tropospheric Effects	± 0.5 m	± 0.5 m
Multipath Distortion Effects	± 1 m	± 1 m
Satellite Clock Errors	± 2 m	± 0.5 m
Ephemeris Errors	± 2.5 m	± 0.3 m
Signal Arrival Errors	± 3 m	± 0.03 m

The standard deviation of the User Equivalent Range Errors, σ_R, is an important figure, and is usually computed from

$$\sigma_R = \sqrt{\sum_{i=1}^{n} (\text{UERE}_i)^2} \tag{5}$$

where UERE $_i$ is the individual source of error. From the table σ_R is found to be equal to 6.7 m in the current system, and 5.1 m in the planned new system.

2.5. Dilution of Precision (DOP)

Since the Position Dilution of Precision (PDOP) is strictly a function of the geometrical distribution of the satellites [8, 12], it is not expected to be different in the new system. In addition, the Geometric Dilution of Precision (GDOP), which factors in the normalized receiver clock bias, is not expected to be substantially different.

[3]The carrier frequency has not been specified as of this writing; however, it is expected to be in the L band, as in the current system.

2.6. The Sagnac effect

The Sagnac effect is an additional timing error that occurs due to the diurnal motion of the receiver during signal propagation, as the earth rotates. As in the current system, the receiver will apply a well known formula [2, 14, 15] to correct for the Sagnac effect.

Conclusion

The third-generation GPS system, now in development by Celestia Corp., has substantial advantages over the existing GPS system. The elimination of the Doppler correction and signal tracking circuitry in the receivers, in addition to the range-independent closed-form solution presented in this article, will allow the size of the receiver to be substantially reduced. In addition, the placement of the satellites in relativistic-null orbits will substantially reduce the amount of maintenance and timing correction uploads to the satellites. Simulations have shown that the User Equivalent Range Errors associated with signal arrival, ephemeris data, and satellite clock drift will be significantly lower in the new system. One final substantial advantage of the new system will be the very small time delay (only a few sec.) required to obtain a first fix on position; since all the ephemeris data will be transmitted from a stationary GEO, on one dedicated digital channel. Designs for new, miniaturized GPS receivers that will work with the third-generation GPS system are currently being planned by a number of receiver manufacturers.

Appendix: Closed-form Solution of the GPS Geolocation Equations

A closed-form solution for the GPS geolocation equations will be very desirable for further miniaturization of the receiver. Closed-form solutions were published in the past [16, 17]; however, those solutions generally require the computation of the range along with the coordinates of the receiver. Unfortunately, when the range data is computed along with the receiver's coordinates, unacceptably large errors in the receiver's coordinates result as a consequence. For that reason, iterative solutions have been devised and continue to be preferred over closed-form solutions [8, 12]. The downside of solving the geolocation equations iteratively, however, is that the circuitry required (microprocessor, memory, etc.) contribute significantly to the large size of the receiver[4]. This chapter provides for the first time a closed-form solution for the GPS geolocation equations that does not require the calculation of range data. The simulation results given in the following section show that the accuracy of the computed coordinates of the receiver matches the accuracy that is obtainable with iterative solutions. The benefit of a closed-form solution is that the solution can be entirely carried out in one miniature VLSI chip[5].

[4]Current GPS receivers typically rely on a 300 MHz microprocessor, at least 64MB of RAM, and at least 64MB of Flash ROM.

[5]A VLSI chip has been designed for implementing the closed-form solution presented here, and measures only 5 mm $\times$ 5 mm. The chip calculates the solution in a fraction of the time that is required in current GPS

The basic principle of the GPS is quite simple. A receiver acquires signals that were transmitted simultaneously from four different satellites that exist at positions $\vec{P}_i$. The signals arrive at the receiver at times t_i; all times and positions being measured in an earth-centric, earth-fixed coordinate system. The receiver's position $\vec{P}$ as well as the initial time of transmission t of the signals[6] can then be determined by solving four simultaneous equations [8, 12]:

$$\left|\vec{P} - \vec{P}_i\right| = c\,(t_i - t); \qquad i = 1, 2, 3, 4, \tag{6}$$

where c is the velocity of light. The set of equations (6) are nonlinear hyperbolic equations; and GPS receivers generally solve those equations iteratively as was pointed out. The closed-form solution that will be now presented is essentially based on transforming the hyperbolic equations (6) into a set of vector equations. It will be then shown that the vector equations are linear and that the position vector of the receiver can be separated from those equations. By squaring Eq.(6) for any index i, we have

$$\left|\vec{P} - \vec{P}_i\right|^2 = c^2\,(t^2 + t_i^2 - 2tt_i) \tag{7}$$

We first note that terms such as $|\vec{P} - \vec{P}_i|^2$ can be represented as follows:

$$\begin{aligned} |\vec{P} - \vec{P}_i|^2 &= (X - X_i)^2 + (Y - Y_i)^2 + (Z - Z_i)^2 \\ &= (X^2 + Y^2 + Z^2) + (X_i^2 + Y_i^2 + Z_i^2) - 2(XX_i + YY_i + ZZ_i) \\ &= |\vec{P}|^2 + |\vec{P}_i|^2 - 2\vec{P}_i^{\,T}\vec{P} \end{aligned} \tag{8}$$

where $\vec{P}_i^{\,T}$ denotes the transpose of the vector $\vec{P}_i$. Hence Eq.(7) is rewritten as

$$|\vec{P}|^2 + |\vec{P}_i|^2 - 2\vec{P}_i^{\,T}\vec{P} = c^2\,(t^2 + t_i^2 - 2tt_i) \tag{9}$$

Writing the above equation explicitly for satellites 1 and 2 (that is, for $i = 1, 2$) and taking the difference between the two resulting equations gives

$$|\vec{P}_1|^2 - |\vec{P}_2|^2 - 2\left(\vec{P}_1^{\,T} - \vec{P}_2^{\,T}\right)\vec{P} = c^2\,\left(t_1^2 - t_2^2 - 2t(t_1 - t_2)\right) \tag{10}$$

Dividing the last equation by $c^2(t_1 - t_2)$ gives

$$\frac{1}{c^2(t_1 - t_2)}\left[|\vec{P}_1|^2 - |\vec{P}_2|^2 - 2\left(\vec{P}_1^{\,T} - \vec{P}_2^{\,T}\right)\vec{P}\right] = (t_1 + t_2) - 2t \tag{11}$$

Eq.(11) is essentially a representation of Eq.(6) in vector form (note that the r.h.s. of the equation is a scalar). Besides the unknown initial time t, the only other unknown in Eq.(11) is the position vector $\vec{P}$. To solve for the coordinates of the receiver, we need to obtain four *linearly independent* equations like Eq.(11). Hence, in total, we must form equations from satellite pairs: 2-1, 3-1, 4-1, and 5-1. This will result in four linearly independent equations. Obviously, we need 5 different satellites to achieve this task. It is therefore

---

receivers.

[6]Note that in GPS the initial time of transmission t is considered to be an unknown due to the lack of perfect synchronization between the receiver's clock and the satellites' clocks.

clear that this solution requires that 5 satellites must be viewable to the receiver, as opposed to 4 satellites in the other solutions. However, making one extra satellite viewable to the receiver is obviously a small price to pay when the reward is substantial miniaturization of the receiver. By following the same analysis outlined above for satellite pairs 3-1, 4-1, and 5-1, we obtain the following three additional equations:

$$\frac{1}{c^2(t_1 - t_3)} \left[|\vec{P}_1|^2 - |\vec{P}_3|^2 - 2\left(\vec{P}_1^{\,T} - \vec{P}_3^{\,T}\right)\vec{P} \right] = (t_1 + t_3) - 2t \qquad (12)$$

$$\frac{1}{c^2(t_1 - t_4)} \left[|\vec{P}_1|^2 - |\vec{P}_4|^2 - 2\left(\vec{P}_1^{\,T} - \vec{P}_4^{\,T}\right)\vec{P} \right] = (t_1 + t_4) - 2t \qquad (13)$$

$$\frac{1}{c^2(t_1 - t_5)} \left[|\vec{P}_1|^2 - |\vec{P}_5|^2 - 2\left(\vec{P}_1^{\,T} - \vec{P}_5^{\,T}\right)\vec{P} \right] = (t_1 + t_5) - 2t \qquad (14)$$

Equations (11), (12), (13), and (14) are the required four simultaneous equations in the unknown coordinates of the receiver $\vec{P}$ and the time of transmission t. Eliminating t from Eqs.(11) and (12) gives

$$\left[\frac{\vec{P}_1^{\,T} - \vec{P}_3^{\,T}}{(t_1 - t_3)} - \frac{\vec{P}_1^{\,T} - \vec{P}_2^{\,T}}{(t_1 - t_2)} \right] \vec{P} = \frac{1}{2} \left[\frac{|\vec{P}_1|^2 - |\vec{P}_3|^2}{(t_1 - t_3)} - \frac{|\vec{P}_1|^2 - |\vec{P}_2|^2}{(t_1 - t_2)} - c^2(t_3 - t_2) \right] \qquad (15)$$

By similarly eliminating t from equation pairs (11),(13) and (11),(14) we get

$$\left[\frac{\vec{P}_1^{\,T} - \vec{P}_4^{\,T}}{(t_1 - t_4)} - \frac{\vec{P}_1^{\,T} - \vec{P}_2^{\,T}}{(t_1 - t_2)} \right] \vec{P} = \frac{1}{2} \left[\frac{|\vec{P}_1|^2 - |\vec{P}_4|^2}{(t_1 - t_4)} - \frac{|\vec{P}_1|^2 - |\vec{P}_2|^2}{(t_1 - t_2)} - c^2(t_4 - t_2) \right] \qquad (16)$$

and

$$\left[\frac{\vec{P}_1^{\,T} - \vec{P}_5^{\,T}}{(t_1 - t_5)} - \frac{\vec{P}_1^{\,T} - \vec{P}_2^{\,T}}{(t_1 - t_2)} \right] \vec{P} = \frac{1}{2} \left[\frac{|\vec{P}_1|^2 - |\vec{P}_5|^2}{(t_1 - t_5)} - \frac{|\vec{P}_1|^2 - |\vec{P}_2|^2}{(t_1 - t_2)} - c^2(t_5 - t_2) \right] \qquad (17)$$

The simultaneous equations (15),(16), and (17) can now be written in the alternative algebraic form

$$\begin{aligned} a_{11}X + a_{12}Y + a_{13}Z &= b_1 \\ a_{21}X + a_{22}Y + a_{23}Z &= b_2 \\ a_{31}X + a_{32}Y + a_{33}Z &= b_3 \end{aligned} \qquad (18)$$

where X, Y, and Z are the coordinates of the receiver. Notice that each of these equations is the equation of a plane. The intersection of three planes defines a single point: the location of the receiver. By inspection of Eq.(15), the constants in the first equation will be given by

$$a_{11} = \frac{X_1 - X_3}{(t_1 - t_3)} - \frac{X_1 - X_2}{(t_1 - t_2)}$$

$$
\begin{aligned}
a_{12} &= \frac{Y_1 - Y_3}{(t_1 - t_3)} - \frac{Y_1 - Y_2}{(t_1 - t_2)} \\
a_{13} &= \frac{Z_1 - Z_3}{(t_1 - t_3)} - \frac{Z_1 - Z_2}{(t_1 - t_2)} \\
b_1 &= \frac{1}{2}\left[\frac{(X_1^2 + Y_1^2 + Z_1^2) - (X_3^2 + Y_3^2 + Z_3^2)}{(t_1 - t_3)}\right. \\
&\quad \left. - \frac{(X_1^2 + Y_1^2 + Z_1^2) - (X_2^2 + Y_2^2 + Z_2^2)}{(t_1 - t_2)} - c^2(t_3 - t_2)\right] \qquad (19)
\end{aligned}
$$

The other constants can be similarly determined from (16) and (17) by inspection. The set of equations (15), (16), and (17) constitute the desired closed-form solution of the GPS geolocation problem. It is not difficult to verify that those equations can be used to compute the receiver's coordinates precisely (by using hypothetical numbers for the coordinates of the satellites and the receiver, calculating the times of arrival, and substituting in the equations, it can be easily verified that the equations reproduce the exact coordinates of the receiver). An important observation is that range data is not present in equations (15), (16), and (17).

References

[1] IEEE Spectrum, *The Big Picture*, March 2008, p. 18.

[2] N. Ashby, *Relativity and the Global Positioning System*, Physics Today, May 2002, p. 41.

[3] N. Ashby, *Relativity and the Global Positioning System*, Living Reviews in Relativity, V. 6, 2003, pp. 1-43.

[4] G.D. Gordon and W.L. Morgan, *Principles of Communication Satellites*, (John Wiley, New York, NY, 1993).

[5] B.D. Tapley, B.E. Schutz, and G.H. Born, *Statistical Orbit Determination*, (Elsevier Academic Press, Burlington, MA, 2004).

[6] W.M. Kaula, *Theory of Satellite Geodesy*, (Dover, New York, NY, 2000).

[7] W. Torge, *Geodesy*, (deGruyter, Berlin, Germany, 2001).

[8] J.B. Tsui, *Fundamentals of Global Positioning System Receivers*, (John Wiley, New York, NY, 2000).

[9] W. Zumberge and I. Bertiger, *Ephemeris and Clock Navigation Message Accuracy*, in *Global Positioning System Theory and Applications*, Vol. I, B.W. Parkinson, Ed. (American Institute of Aeronautics and Astronautics, Cambridge, MA, 1996).

[10] Lockheed Martin Federal Systems, *Global Positioning System Operational Control Segment Performance Analysis - Final Report*, (available by request from the NAVSTAR GPS Joint Program Office, CZG), Sep. 30, 1996.

[11] S.C. Fisher and K. Ghassemi, *GPS IIF - The Next Generation*, Proceedings of the IEEE, 87, 1, 1999, pp. 24-47.

[12] M.S. Grewal, L.R. Weill, and A.P. Andrews, *Global Positioning Systems, Inertial Navigation, and Integration*, (John Wiley, New York, NY, 2001).

[13] D. Acosta et. al, *A Time-of-Flight Detector in CDF-II*, Nuclear Instruments and Methods A, 518, 1-2, 2004, pp. 605-608.

[14] L. Landau and E. Lifshitz, *The Classical Theory of Fields*, (Pergamon Press, New York, NY, 1997).

[15] D. Allan, N. Ashby, and M. Weiss, *Around-the-World Relativistic Sagnac Experiment*, Science, 228, 4695, 1985, pp. 69-70.

[16] J. Hoshen, *The GPS equations and the Problem of Apollonius*, IEEE Trans. on Aerospace and Electronic Systems, 32, 3, 1996, pp. 1116-1124.

[17] J.L. Leva, *An alternative closed-form solution to the GPS pseudo-range equations*, IEEE Trans. on Aerospace and Electronic Systems, 32, 4, 1996, pp. 1430-1439.

INDEX

M

N

O

P

T

U

V

W

Y